ASP Prep Book

Study Guide with Practice Test Questions for the Associate Safety Professional Certification Exam [Includes Detailed Answer Explanations]

Joshua Rueda

Written and edited by TPB Publishing.

ISBN 13: 9781637750650
ISBN 10: 163775065X

Table of Contents

Quick Overview 1

Test-Taking Strategies 2

Introduction to the ASP Exam 6

Study Prep Plan for the ASP Exam 8

Advanced Sciences and Math 10

General Chemistry Concepts 10

Electrical Principles 13

Principles of Radioactivity 21

Storage Capacity Calculations 25

Rigging and Load Calculations 25

Ventilation and System Design 26

Noise Hazards 27

Climate and Environmental Conditions 28

Fall Protection Calculations 30

General Physics Concepts 31

Financial Principles 36

Descriptive Statistics 38

Lagging Indicators 45

Leading Indicators 47

Safety Management Systems 49

Hierarchy of Hazard Controls 49

Risk Transfer 52

Management of Change 54

Hazard and Risk Analysis Methods 55

Process Safety Management 60
Fleet Safety Principles 60
Hazard Communication and Globally Harmonized System 64
Control of Hazardous Energy 65
Excavation, Trenching, and Shoring 66
Confined Space 68
Physical Security 69
Fall Protection 69
Machine Guarding 71
Powered Industrial Vehicles 71
Scaffolding 72
Using Hazard Identification Methods 74
Assessing and Analyzing Risks 76
Providing Financial Justification of Hazard Controls 78
Implementing Hazard Controls 79
Monitoring and Reevaluating Hazard Controls 81
Conducting Incident Investigations 84
Conducting Inspections and Audits 87
Evaluating Cost, Schedule, Performance, and Project Risk 89

Ergonomics 90

Fitness for Duty 90
Stressors 90
Risk Factors 91
Work Design 92
Material Handling 93
Work Practice Controls 95

Using Qualitative and Quantitative Analysis Methods 96

Fire Prevention and Protection........ 99

Chemical Hazards........ 99

Electrical Hazards 99

Hot Work........ 103

Fire Science and Combustible Dust 104

Detection Systems 105

Suppression Systems, Fire Extinguishers, and Sprinkler Types........ 106

Segregation and Separation 109

Housekeeping 110

Emergency Response Management (ERM) 111

Emergency, Crisis, Disaster Response Planning........ 111

Workplace Violence 114

Industry Hygiene and Occupational Health........ 116

Sources of Biological Hazards 116

Protocol for Bloodborne Pathogen Control........ 118

Mutagens, Teratogens, and Carcinogens 119

Chemical Hazards........ 120

Exposure Limits 122

Routes of Entry........ 125

Acute and Chronic Exposures........ 126

Noise........ 128

Radiation 129

Heat and Cold Stress 131

Conducting an Exposure Assessment 131

Environmental Management 133

Environmental Hazards Awareness 133
Water 134
Air 135
Land and Conservation 136
Hierarchy of Conservation 137
Environmental Management System Standards 139
Waste Removal, Treatment, and Disposal 140
Training, Education, and Communication 144
Adult Learning Theory and Techniques 146
Presentation Tools 147
Safety Culture/Climate 148
Data Collection, Needs Analysis, Gap Analysis, and Feedback 150
Assessing Competency 153
Law and Ethics 154
Legal Liability 154
Ethical Behavior 155
Protection of Worker Privacy 158
Dealing with Unethical Situations 159
Reading and Interpreting Regulations 160
Determining Appropriate Actions Based on Knowledge Limitations 161
Practice Test 1
Answer Explanations 32
Index 61

Quick Overview

As you draw closer to taking your exam, effective preparation becomes more and more important. Thankfully, you have this study guide to help you get ready. Use this guide to help keep your studying on track and refer to it often.

This study guide contains several key sections that will help you be successful on your exam. The guide contains tips for what you should do the night before and the day of the test. Also included are test-taking tips. Knowing the right information is not always enough. Many well-prepared test takers struggle with exams. These tips will help equip you to accurately read, assess, and answer test questions.

A large part of the guide is devoted to showing you what content to expect on the exam and to helping you better understand that content. In this guide are practice test questions so that you can see how well you have grasped the content. Then, answer explanations are provided so that you can understand why you missed certain questions.

Don't try to cram the night before you take your exam. This is not a wise strategy for a few reasons. First, your retention of the information will be low. Your time would be better used by reviewing information you already know rather than trying to learn a lot of new information. Second, you will likely become stressed as you try to gain a large amount of knowledge in a short amount of time. Third, you will be depriving yourself of sleep. So be sure to go to bed at a reasonable time the night before. Being well-rested helps you focus and remain calm.

Be sure to eat a substantial breakfast the morning of the exam. If you are taking the exam in the afternoon, be sure to have a good lunch as well. Being hungry is distracting and can make it difficult to focus. You have hopefully spent lots of time preparing for the exam. Don't let an empty stomach get in the way of success!

When travelling to the testing center, leave earlier than needed. That way, you have a buffer in case you experience any delays. This will help you remain calm and will keep you from missing your appointment time at the testing center.

Be sure to pace yourself during the exam. Don't try to rush through the exam. There is no need to risk performing poorly on the exam just so you can leave the testing center early. Allow yourself to use all of the allotted time if needed.

Remain positive while taking the exam even if you feel like you are performing poorly. Thinking about the content you should have mastered will not help you perform better on the exam.

Once the exam is complete, take some time to relax. Even if you feel that you need to take the exam again, you will be well served by some down time before you begin studying again. It's often easier to convince yourself to study if you know that it will come with a reward!

Test-Taking Strategies

1. Predicting the Answer

When you feel confident in your preparation for a multiple-choice test, try predicting the answer before reading the answer choices. This is especially useful on questions that test objective factual knowledge. By predicting the answer before reading the available choices, you eliminate the possibility that you will be distracted or led astray by an incorrect answer choice. You will feel more confident in your selection if you read the question, predict the answer, and then find your prediction among the answer choices. After using this strategy, be sure to still read all of the answer choices carefully and completely. If you feel unprepared, you should not attempt to predict the answers. This would be a waste of time and an opportunity for your mind to wander in the wrong direction.

2. Reading the Whole Question

Too often, test takers scan a multiple-choice question, recognize a few familiar words, and immediately jump to the answer choices. Test authors are aware of this common impatience, and they will sometimes prey upon it. For instance, a test author might subtly turn the question into a negative, or he or she might redirect the focus of the question right at the end. The only way to avoid falling into these traps is to read the entirety of the question carefully before reading the answer choices.

3. Looking for Wrong Answers

Long and complicated multiple-choice questions can be intimidating. One way to simplify a difficult multiple-choice question is to eliminate all of the answer choices that are clearly wrong. In most sets of answers, there will be at least one selection that can be dismissed right away. If the test is administered on paper, the test taker could draw a line through it to indicate that it may be ignored; otherwise, the test taker will have to perform this operation mentally or on scratch paper. In either case, once the obviously incorrect answers have been eliminated, the remaining choices may be considered. Sometimes identifying the clearly wrong answers will give the test taker some information about the correct answer. For instance, if one of the remaining answer choices is a direct opposite of one of the eliminated answer choices, it may well be the correct answer. The opposite of obviously wrong is obviously right! Of course, this is not always the case. Some answers are obviously incorrect simply because they are irrelevant to the question being asked. Still, identifying and eliminating some incorrect answer choices is a good way to simplify a multiple-choice question.

4. Don't Overanalyze

Anxious test takers often overanalyze questions. When you are nervous, your brain will often run wild, causing you to make associations and discover clues that don't actually exist. If you feel that this may be a problem for you, do whatever you can to slow down during the test. Try taking a deep breath or counting to ten. As you read and consider the question, restrict yourself to the particular words used by the author. Avoid thought tangents about what the author *really* meant, or what he or she was *trying* to say. The only things that matter on a multiple-choice test are the words that are actually in the question. You must avoid reading too much into a multiple-choice question, or supposing that the writer meant something other than what he or she wrote.

5. No Need for Panic

It is wise to learn as many strategies as possible before taking a multiple-choice test, but it is likely that you will come across a few questions for which you simply don't know the answer. In this situation, avoid panicking. Because most multiple-choice tests include dozens of questions, the relative value of a single wrong answer is small. As much as possible, you should compartmentalize each question on a multiple-choice test. In other words, you should not allow your feelings about one question to affect your success on the others. When you find a question that you either don't understand or don't know how to answer, just take a deep breath and do your best. Read the entire question slowly and carefully. Try rephrasing the question a couple of different ways. Then, read all of the answer choices carefully. After eliminating obviously wrong answers, make a selection and move on to the next question.

6. Confusing Answer Choices

When working on a difficult multiple-choice question, there may be a tendency to focus on the answer choices that are the easiest to understand. Many people, whether consciously or not, gravitate to the answer choices that require the least concentration, knowledge, and memory. This is a mistake. When you come across an answer choice that is confusing, you should give it extra attention. A question might be confusing because you do not know the subject matter to which it refers. If this is the case, don't eliminate the answer before you have affirmatively settled on another. When you come across an answer choice of this type, set it aside as you look at the remaining choices. If you can confidently assert that one of the other choices is correct, you can leave the confusing answer aside. Otherwise, you will need to take a moment to try to better understand the confusing answer choice. Rephrasing is one way to tease out the sense of a confusing answer choice.

7. Your First Instinct

Many people struggle with multiple-choice tests because they overthink the questions. If you have studied sufficiently for the test, you should be prepared to trust your first instinct once you have carefully and completely read the question and all of the answer choices. There is a great deal of research suggesting that the mind can come to the correct conclusion very quickly once it has obtained all of the relevant information. At times, it may seem to you as if your intuition is working faster even than your reasoning mind. This may in fact be true. The knowledge you obtain while studying may be retrieved from your subconscious before you have a chance to work out the associations that support it. Verify your instinct by working out the reasons that it should be trusted.

8. Key Words

Many test takers struggle with multiple-choice questions because they have poor reading comprehension skills. Quickly reading and understanding a multiple-choice question requires a mixture of skill and experience. To help with this, try jotting down a few key words and phrases on a piece of scrap paper. Doing this concentrates the process of reading and forces the mind to weigh the relative importance of the question's parts. In selecting words and phrases to write down, the test taker thinks about the question more deeply and carefully. This is especially true for multiple-choice questions that are preceded by a long prompt.

9. Subtle Negatives

One of the oldest tricks in the multiple-choice test writer's book is to subtly reverse the meaning of a question with a word like *not* or *except*. If you are not paying attention to each word in the question, you can easily be led astray by this trick. For instance, a common question format is, "Which of the following is...?" Obviously, if the question instead is, "Which of the following is not...?," then the answer will be quite different. Even worse, the test makers are aware of the potential for this mistake and will include one answer choice that would be correct if the question were not negated or reversed. A test taker who misses the reversal will find what he or she believes to be a correct answer and will be so confident that he or she will fail to reread the question and discover the original error. The only way to avoid this is to practice a wide variety of multiple-choice questions and to pay close attention to each and every word.

10. Reading Every Answer Choice

It may seem obvious, but you should always read every one of the answer choices! Too many test takers fall into the habit of scanning the question and assuming that they understand the question because they recognize a few key words. From there, they pick the first answer choice that answers the question they believe they have read. Test takers who read all of the answer choices might discover that one of the latter answer choices is actually *more* correct. Moreover, reading all of the answer choices can remind you of facts related to the question that can help you arrive at the correct answer. Sometimes, a misstatement or incorrect detail in one of the latter answer choices will trigger your memory of the subject and will enable you to find the right answer. Failing to read all of the answer choices is like not reading all of the items on a restaurant menu: you might miss out on the perfect choice.

11. Spot the Hedges

One of the keys to success on multiple-choice tests is paying close attention to every word. This is never truer than with words like almost, most, some, and sometimes. These words are called "hedges" because they indicate that a statement is not totally true or not true in every place and time. An absolute statement will contain no hedges, but in many subjects, the answers are not always straightforward or absolute. There are always exceptions to the rules in these subjects. For this reason, you should favor those multiple-choice questions that contain hedging language. The presence of qualifying words indicates that the author is taking special care with their words, which is certainly important when composing the right answer. After all, there are many ways to be wrong, but there is only one way to be right! For this reason, it is wise to avoid answers that are absolute when taking a multiple-choice test. An absolute answer is one that says things are either all one way or all another. They often include words like *every*, *always*, *best*, and *never*. If you are taking a multiple-choice test in a subject that doesn't lend itself to absolute answers, be on your guard if you see any of these words.

12. Long Answers

In many subject areas, the answers are not simple. As already mentioned, the right answer often requires hedges. Another common feature of the answers to a complex or subjective question are qualifying clauses, which are groups of words that subtly modify the meaning of the sentence. If the question or answer choice describes a rule to which there are exceptions or the subject matter is complicated, ambiguous, or confusing, the correct answer will require many words in order to be expressed clearly and accurately. In essence, you should not be deterred by answer choices that seem excessively long. Oftentimes, the author of the text will not be able to write the correct answer without offering some qualifications and modifications. Your job is to read the answer choices thoroughly and completely and to select the one that most accurately and precisely answers the question.

13. Restating to Understand

Sometimes, a question on a multiple-choice test is difficult not because of what it asks but because of how it is written. If this is the case, restate the question or answer choice in different words. This process serves a couple of important purposes. First, it forces you to concentrate on the core of the question. In order to rephrase the question accurately, you have to understand it well. Rephrasing the question will concentrate your mind on the key words and ideas. Second, it will present the information to your mind in a fresh way. This process may trigger your memory and render some useful scrap of information picked up while studying.

14. True Statements

Sometimes an answer choice will be true in itself, but it does not answer the question. This is one of the main reasons why it is essential to read the question carefully and completely before proceeding to the answer choices. Too often, test takers skip ahead to the answer choices and look for true statements. Having found one of these, they are content to select it without reference to the question above. Obviously, this provides an easy way for test makers to play tricks. The savvy test taker will always read the entire question before turning to the answer choices. Then, having settled on a correct answer choice, he or she will refer to the original question and ensure that the selected answer is relevant. The mistake of choosing a correct-but-irrelevant answer choice is especially common on questions related to specific pieces of objective knowledge. A prepared test taker will have a wealth of factual knowledge at their disposal, and should not be careless in its application.

15. No Patterns

One of the more dangerous ideas that circulates about multiple-choice tests is that the correct answers tend to fall into patterns. These erroneous ideas range from a belief that B and C are the most common right answers, to the idea that an unprepared test-taker should answer "A-B-A-C-A-D-A-B-A." It cannot be emphasized enough that pattern-seeking of this type is exactly the WRONG way to approach a multiple-choice test. To begin with, it is highly unlikely that the test maker will plot the correct answers according to some predetermined pattern. The questions are scrambled and delivered in a random order. Furthermore, even if the test maker was following a pattern in the assignation of correct answers, there is no reason why the test taker would know which pattern he or she was using. Any attempt to discern a pattern in the answer choices is a waste of time and a distraction from the real work of taking the test. A test taker would be much better served by extra preparation before the test than by reliance on a pattern in the answers.

Introduction to the ASP Exam

Function of the Test

The Associate Safety Professional (ASP) certification is a safety, health, and environment credential offered by the Board of Certified Safety Professionals (BCSP). The ASP certification is often listed as a preferred or required qualification in job postings for safety professionals, so obtaining it can qualify a candidate for jobs or help them stand out in the hiring process. This credential demonstrates mastery of professional safety knowledge as well as offers a competitive advantage when professionals are seeking assignments, promotions, or salary increases. Additionally, earning the ASP certification is one step required to earn the more comprehensive BCSP Certified Safety Professional (CSP) certification.

To be eligible for the ASP credential, an individual must either have a minimum of a bachelor's degree in any field or an associate degree in safety, health, or the environment. The associate degree must include at least four courses with at least 12 semester hours/18 quarter hours of study in the safety, health, or environmental domains covered in the ASP examination blueprint. Degrees earned outside of the United States will be evaluated for equivalency. The candidate must also have at least one year of safety experience where safety is at least 50% preventative, professional level with a wide range of safety duties, such as evaluating risks, investigating incidents, and managing hazardous materials. If the individual meets these criteria, then they must pass the ASP examination to earn certification.

Once the certification is earned, professionals can retain certification by paying an annual renewal fee of $170.00 and meeting recertification requirements. A prorated renewal fee is applied for the remainder of the year once a candidate passes an exam. Finally, credential holders must also earn and submit 25 recertification points every five years to keep the certification.

Test Administration

Candidates should create a profile and apply for the ASP credential through the BCSP certification website at www.bcsp.org. The application requires contact information, experience information, education and training information, an application agreement and validation, and payment. The application fee is $160.00; however, application fee waivers are available to government employees, college or university faculty that teach safety-related courses, and U.S. military members and veterans.

Once the application has been approved, candidates receive their Examination Authorization Letter, which lists the Candidate ID needed to schedule the exam. Candidates then have an eligibility period of one year to schedule their exam online at www.pearsonvue.com/bcsp or by phone. BCSP recommends scheduling the exam at least six to eight weeks prior to the test date. The individual exam fee is $350.00. Within the last 60 days of eligibility, candidates can purchase an additional year of eligibility. However, eligibility can only be extended once per term. If the candidate does not complete the test during the initial term or during an extension, then they must reapply.

The test is administered via computer at any Pearson VUE test center the candidate elects. On the day of the test, individuals should bring a valid, unexpired government issued photo ID with the candidate's signature. Materials for working out solutions by hand are provided by the testing center. An on-screen calculator, which is designed to mimic the TI-30XS scientific calculator, is provided during the exam. The candidate is not allowed to bring any personal belongings into the testing room or access outside references during the exam.

Reasonable accommodations are provided in accordance with the American Disabilities Act and the American Disabilities Amendment Act; however, the candidate must request accommodations when purchasing the exam by emailing accommodations@bcsp.org

Test Format

The test comprises 200 multiple-choice questions, each with 4 possible answers and one correct answer choice. A certain number of the questions (10 – 15%) are experimental and, thus, do not influence a candidate's final score. The test-taker is given 5 hours to complete the ASP exam. They may mark items to return to later or skip questions during the test. This information is displayed in a table at the end of the exam, prior to submission, so the candidate may return to any marked or skipped items.

While the test is not divided into sections, BCSP categorizes the test content into nine domains. Below, each domain is listed along with the percentage of corresponding questions present on the exam:

Domain	Test Content	Percentage
Domain 1	Advanced Sciences and Math	11.55%
Domain 2	Safety Management Systems	17.22%
Domain 3	Ergonomics	9.00%
Domain 4	Fire Prevention and Protection	10.66%
Domain 5	Emergency Response Management (ERM)	9.57%
Domain 6	Industrial Hygiene and Occupational Health	12.59%
Domain 7	Environmental Management	8.68%
Domain 8	Training, Education, and Communication	12.35%
Domain 9	Law and Ethics	8.38%

Scoring

When the candidate submits the exam, they immediately receive a pass/fail result. A minimum passing score determines whether the candidate obtains the ASP credential. The exact minimum passing score is adjusted for each exam based on the difficulty of the test questions; this ensures equivalency between exams. The minimum passing score corresponds to an individual's total score, and there are no requirements to obtain a certain score in any specific domain.

The pass rate for the ASP examination is approximately 50%. As of July 2, 2018, if the candidate does not pass, they may purchase an authorization to retake the exam. They are allowed to retake the ASP test beginning six weeks after the most recent attempt if their eligibility remains intact.

Study Prep Plan for the ASP Exam

1 **Schedule -** Use one of our study schedules below or come up with one of your own.

2 **Relax -** Test anxiety can hurt even the best students. There are many ways to reduce stress. Find the one that works best for you.

3 **Execute -** Once you have a good plan in place, be sure to stick to it.

One Week Study Schedule

Day 1	Advanced Sciences and Math
Day 2	Safety Management Systems
Day 3	Ergonomics
Day 4	Industry Hygiene and Occupational Health
Day 5	Training, Education, and Communication
Day 6	Practice Questions
Day 7	Take Your Exam!

Two Week Study Schedule

Day 1	Advanced Sciences and Math	Day 8	Emergency Response Management (ERM)
Day 2	Storage Capacity Calculations	Day 9	Environmental Management
Day 3	Descriptive Statistics	Day 10	Training, Education...
Day 4	Safety Management Systems	Day 11	Law and Ethics
Day 5	Hazard Communication and Globally...	Day 12	Practice Questions
Day 6	Implementing Hazard Controls	Day 13	Answer Explanations
Day 7	Ergonomics	Day 14	Take Your Exam!

One Month Study Schedule

Day 1	Advanced Sciences and Math	Day 11	Excavation, Trenching, and...	Day 21	Exposure Limits
Day 2	Electrical Principles	Day 12	Machine Guarding	Day 22	Noise
Day 3	Principles of Radioactivity	Day 13	Implementing Hazard Controls	Day 23	Environmental Management
Day 4	Noise Hazards	Day 14	Conducting Incident Investigations	Day 24	Environmental Management System Standards
Day 5	General Physics Concepts	Day 15	Ergonomics	Day 25	Adult Learning Theory and Techniques
Day 6	Descriptive Statistics	Day 16	Material Handling	Day 26	Assessing Competency
Day 7	Lagging Indicators	Day 17	Fire Prevention and Protection	Day 27	Protection of Worker Privacy
Day 8	Safety Management Systems	Day 18	Detection Systems	Day 28	Practice Questions
Day 9	Hazard and Risk Analysis Methods	Day 19	Emergency Response Management (ERM)	Day 29	Answer Explanations
Day 10	Fleet Safety Principles	Day 20	Industry Hygiene and Occupational Health	Day 30	Take Your Exam!

Advanced Sciences and Math

General Chemistry Concepts

Nomenclature

The word chemical has a broad definition. It may refer to elements, compounds, or a combination of elements or compounds. Chemicals are given chemical names that conform with either the **International Union of Pure and Applied Chemistry (IUPAC)** or **Chemical Abstract Service (CAS)** nomenclature system. Each entity has its own rules of nomenclature under their respective systems. However, IUPAC is generally the preferred system. The purpose of the nomenclature systems is to designate a name that unambiguously determines the chemical for hazard evaluations.

The IUPAC nomenclature uses the following for naming chemicals:

Suffix – determines the functional group that is held within the compound. The classification of the functional group determines the name of the suffix.

- alkane: -ane
- alkene: -ene
- alkyne: -yne
- alcohol: -ol
- aldehyde: -al
- ketone: -one
- carboxylic acid: -oic acid
- ester: -oate

Prefix – locate the longest carbon chain that holds the functional group. The number of carbon atoms located within that chain determines the prefix of the compound.

- 1: meth-
- 2: eth-
- 3: prop-
- 4: but-
- 5: pent-
- 6: hex-
- 7: hept-
- 8: oct-
- 9: non-
- 10: dec-

Next, a numbering is designated based on the number of carbons on the longest carbon chain, but the chemical's functioning group will show on the lowest possible carbon. Generally, numbering begins at the carbon that is nearest to the functional group. Branched groups are named by the number of carbon atoms on a branch. If there are multiple branch groups that are the same type, then two numbers will be designated. However, if it is an alkane, then the carbon with the lowest number must hold the branched group.

Regarding alkyl halides, the halogen atom is regarded in a similar manner as branched groups. Halogen atom naming requires converting the halogen atom (e.g. bromine) by changing the ending -ine to -o (e.g. bromo).

fluorine → fluoro
chlorine → chloro
bromine → bromo
iodine → iodo

Halogen atoms may be named based on the location in relation to the carbon chain. The atoms are listed before the main chain. However, this portion may be skipped if there are no halogen atoms.

Finally, the components are combined to produce a single word. The order of the name will be as follows. First are branched groups/halogen atoms (in alphabetical order, but excluding applicable prefixes). Next, the prefix of the main chain is followed by the suffix based on the functional group and longest carbon chain.

Balancing Chemical Equations

Chemical reactions are conveyed using chemical equations. Chemical equations must be balanced with equivalent numbers of atoms for each type of element on each side of the equation. Antoine Lavoisier, a French chemist, was the first to propose the **Law of Conservation of Mass** for the purpose of balancing a chemical equation. The law states, "Matter is neither created nor destroyed during a chemical reaction."

The **reactants** are located on the left side of the arrow, while the **products** are located on the right side of the arrow. Coefficients are the numbers in front of the chemical formulas. Subscripts are the numbers to the lower right of chemical symbols in a formula. To tally atoms, one should multiply the formula's coefficient by the subscript of each chemical symbol. For example, the chemical equation $2\ \mathrm{H_2} + 02 \rightarrow 2\mathrm{H_2O}$ is balanced. For H, the coefficient of 2 multiplied by the subscript 2 equals 4 hydrogen atoms. For O, the coefficient of 1 multiplied by the subscript 2 equals 2 oxygen atoms. Coefficients and subscripts of 1 are understood and never written.

The equations help to explain how the molecules change during a reaction. For example, when hydrogen gas (H_2) combines with oxygen gas (O_2), two molecules of water are formed. The equation is written as follows, where the plus sign means *reacts with* and the arrow means *produces*:

$$2\ \mathrm{H_2} + \mathrm{O_2} \rightarrow 2\ \mathrm{H_2O}$$

Two hydrogen molecules react with an oxygen molecule to produce 2 water molecules. In all chemical equations, the quantity of each element on the reactant side of the equation should equal the quantity of the same element on the product side of the equation. This is due to the law of conservation of matter. If this is true of the equation, the equation is described as balanced. To appropriately label and balance the number of elements on each side of the equation, the coefficient of the element should be multiplied by the subscript next to the element. Coefficients and subscripts are used for quantities larger than one. The **coefficient** is the number located directly to the left of the element.

The **subscript** is the small-sized number directly to the right of the element. In the equation above, on the left side, the coefficient of the hydrogen is 2, and the subscript is also 2, which makes a total of 4 hydrogen atoms. There are 2 oxygen atoms on the left side, so a coefficient of 2 is added in front of water (H_2O) to indicate that there are 2 oxygen atoms. The coefficient multiplied by the subscript in each element of the water molecule also gives 4 hydrogen atoms. This equation is balanced because there are 4 hydrogen atoms and 2 oxygen atoms on each side. The states of the reactants and products can also be written in the equation: gas (g), liquid (l), solid (s), and dissolved in water (aq). If they are included, they are noted in parentheses on the right side of each molecule in the equation.

Chemical Reactions

Chemical reactions are characterized by a chemical change in which the starting substances, or reactants, differ from the substances formed, or products. Chemical reactions may involve a change in color, the production of gas, the formation of a precipitate, or changes in heat content.

The following are the basic types of chemical reactions:

- **Decomposition Reactions:** A compound is broken down into smaller elements. For example, $2H_2O \rightarrow 2H_2 + O_2$. This is read as, "2 molecules of water decompose into 2 molecules of hydrogen and 1 molecule of oxygen."

- **Synthesis Reactions:** Two or more elements or compounds are joined together. For example, $2H_2 + O_2 \rightarrow 2H_2O$. This is read as, "2 molecules of hydrogen react with 1 molecule of oxygen to produce 2 molecules of water."

- **Single Displacement Reactions:** A single element or ion takes the place of another element in a compound. It is also known as a substitution reaction. For example, $Zn + 2\ HCl \rightarrow ZnCl_2 + H_2$. This is read as, "zinc reacts with 2 molecules of hydrochloric acid to produce one molecule of zinc chloride and one molecule of hydrogen." In other words, zinc replaces the hydrogen in hydrochloric acid.

- **Double Displacement Reactions:** Two elements or ions exchange a single element to form two different compounds, resulting in different combinations of cations and anions in the final compounds. It is also known as a metathesis reaction. For example,

  $$H_2SO_4 + 2\ NaOH \rightarrow Na_2SO_4 + 2\ H_2O$$

  - Special types of double displacement reactions include:
    - **Oxidation-Reduction (or Redox) Reactions:** Elements undergo a change in oxidation number. For example,

      $$2\ S_2O_3{}^{2-}(aq) + I_2\ (aq) \rightarrow S_4O_6{}^{2-}(aq) + 2\ I^-\ (aq)$$

    - **Acid-Base Reactions:** Involves a reaction between an acid and a base, which produces a salt and water. For example,

      $$HBr + NaOH \rightarrow NaBr + H_2O$$

    - **Combustion Reactions:** A hydrocarbon (a compound composed of only hydrogen and carbon) reacts with oxygen (O_2) to form carbon dioxide (CO_2) and water (H_2O). For example,

      $$CH_4 + 2O_2 \rightarrow CO_2 + 2H_2O$$

Ideal Gas Law

When different ideal gases are mixed together, the pressure exerted by each gas is independent of the other gases. **Dalton's law** states that the total pressure of a mixture of gases is equal to the sum of the partial pressures of each individual gas and can be calculated with the equation $P_{Total} = P_A + P_B + P_C$. The ideal gas equation, $PV = nRT$, can be used to mathematically relate the pressure (P), volume (V), number of moles (n), and temperature (T) of ideal gases to each other. R is the ideal gas constant, which is equal to 8.314 J mol^{-1} K^{-1}. The pressure and volume of a gas are inversely proportional.

When the pressure of a gas increases, the volume decreases and vice versa. Therefore, according to **Boyle's law**, the left side of the equation, PV, for a gas is always constant. If one property changes, for example, the pressure increases, the volume would decrease and still be equivalent to the product of the original pressure and volume. **Charles's law** states that the relationship between volume and temperature of a gas are directly proportional. When volume is divided by the temperature in Kelvin, the quotient is constant for each gas. When temperature increases, so does the volume of the gas and vice versa. **Avogadro's law** describes the relationship between moles of a gas and the volume of the gas as being directly proportional. As the number of moles increases, the volume also increases. The relationship is constant, similar to that of volume and temperature, so the changes that occur between moles and volume are of the same proportionate magnitude.

pH

Chemical solutions are homogenous mixtures that contain two or more substances. The solute is the less concentrated substance that is dissolved in another substance called the **solvent.** Water is the solvent in an aqueous solution. Chemical solutions may be acidic or basic and can be categorized by the **pH scale**. The pH scale is logarithmic and ranges from 0 to 14. Pure water has a pH of 7, which is considered neutral due to the equal molar concentration of the proton ion (H+) and hydroxide ion (OH–).

Solutions that have a $\mathrm{pH} < 7$ are acidic, and solutions with a $\mathrm{pH} > 7$ are alkaline, or basic. Therefore, the strongest acids have a pH close to 0, and the strongest bases have a pH close to 14. Strong acids, such as hydrochloric acid (HCl), dissolve to produce proton ions ($[\mathrm{H^+}]$), also called hydronium ions $[\mathrm{H_3O^+}]$, in water. The greater the hydrogen ion concentration, the more acidic the solution. The pH is calculated as follows.

$$\mathrm{pH} = -\log_{10}[\mathrm{H^+}]$$

The proton or hydronium concentration is indicated by $[\mathrm{H^+}]$. The brackets indicate the solution concentration in moles per liter (mol/L or M). If the proton concentration is $[\mathrm{H^+}] = 1.00 \times 10^{-5}$ M:

$$\mathrm{pH} = -\log[\mathrm{H^+}] = -\log[1.00 \times 10^{-5}] = 5.000$$

Electrical Principles

Ohm's Law

Ohm's Law describes the relationship between voltage, current, and resistance, which are criteria used to characterize a given circuit. The difference in electrical potential (or voltage drop) between two different points in a circuit can be calculated by multiplying the current between the two points (I) and the total resistance of the electrical devices in the circuit between the two points (R).

$\Delta\text{Voltage } (V) = \text{current } (I) \times \text{resistance } (R)$, where V is voltage (in volts), I is current (in amperes), and R is resistance (in ohms).

Mathematically, the above equation can be manipulated to isolate current, which then is found to be directly proportional to the voltage drop (the electric potential difference) and inversely proportional to the total resistance.

$$I = \frac{V}{R} \text{ or } R = \frac{V}{I}$$

This means that a greater battery voltage (electric potential difference) yields a greater current in the circuit, while greater resistance decreases current. Essentially, charge flows fastest when the battery voltage increases and resistance decreases.

The relationships in these equations can be understood by examining a simple circuit as a reference and then changing one variable sequentially to examine the outcome.

The Relationships Between Voltage, Resistance, and Current in a Basic Circuit

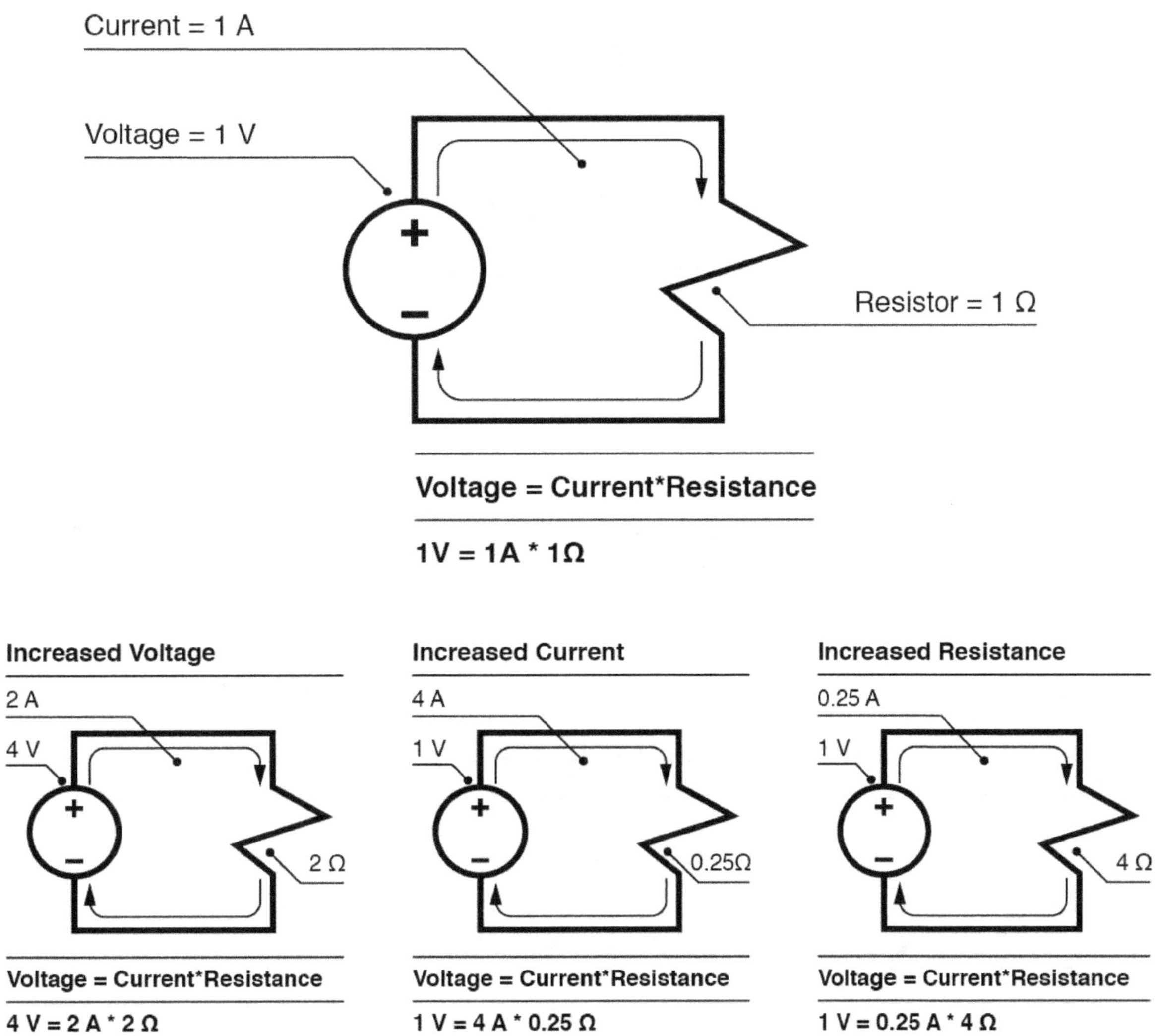

Alternatively, the following “Ohm’s Triangle” is a useful tool to memorize the relationships governed by Ohm’s Law. Test takers will need to memorize this equation for the exam. The triangle serves as a pictorial reminder and method to generate the correct relationships between voltage, current, and resistance.

Ohm’s Triangle

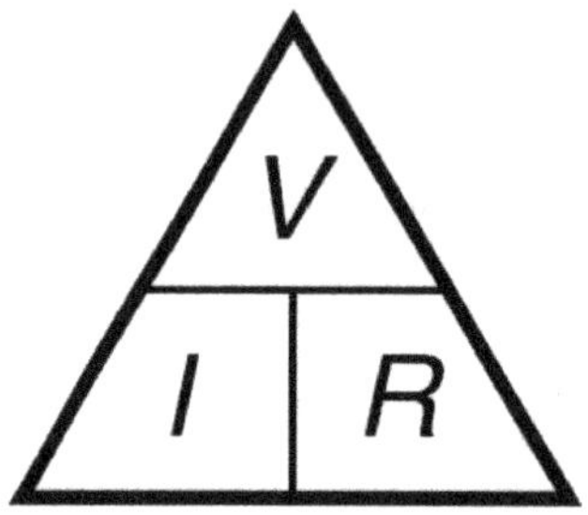

$$V = I \times R$$

Recalling the standard Ohm’s Law that $V = I \times R$ helps set up the basic triangle from which the other two equations can be visually transposed for those who find mathematically manipulating equations difficult.

Ohm’s Triangle:
Voltage, Current, and Resistance Relationships

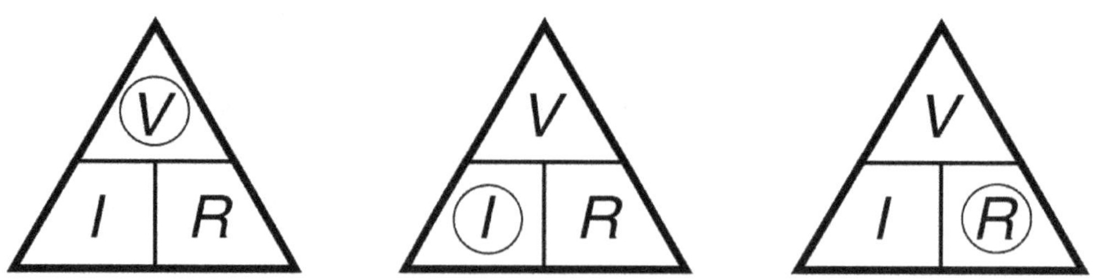

$$V = I \times R \qquad I = \frac{V}{R} \qquad R = \frac{V}{I}$$

Power

Electrical power in a circuit refers to the energy produced or absorbed, and it is expressed in watts (W). In a given circuit, certain components such as light bulbs consume electrical power and convert it to light/heat, while other components, for example, the battery, produce power. Power, in watts, is equal to the current from a voltage source in amperes multiplied by the voltage of that source:

$$\text{Power } (W) = \text{current } (I) \times \text{voltage } (V)$$

Using Ohm's Law and substitution, this relation can also be written in the following two ways:

$$W = I^2R$$

$$W = \frac{V^2}{R}$$

Of note, sometimes power ratings (the rate that electric power is converted to a different form of usable energy [such as light or motion]) on electric motors or other devices is expressed in horsepower, or hp, where $1 \text{ hp} = 746 \text{ W}$.

Impedance

When a voltage is present in an electrical circuit, the opposition the circuit presents to a current is called **impedance (Z)**. Voltage, measured in volts (V), is the electrical force energy that pushes the current between two points. Current, measured in amperes (A), is the electron flow between two points; the current will travel from high to low voltage. Impedance has both a phase and a magnitude, whereas resistance has only magnitude. Resistance, measured in ohms (Ω), is the opposition to electricity flow. The **reactance (X)** is a measure of opposition that a circuit gives, from an inductor or capacitor, to an electrical current when the current is alternating. The units for reactance are ohms (Ω) and include the inductive reactance (X_L) and the capacitive reactance (X_C).

The frequency of the circuit will determine the impedance's opposition to a current (I). The value of the impedance is given by the following equation:

$$Z = \sqrt{R^2 + (X_L - X_C)^2}$$

$Z = \text{impedance (ohms)};\ R = \text{resistance (ohms)}$

Inductive reactance (Ω): $X_L = 2\pi fL$

Capacitive reactance (Ω): $X_C = \frac{1}{2\pi fC}$

Angular frequency: $\omega = 2\pi f$ (radians/second[rad/s])

$f = \text{frequency of oscillating potential (hertz [Hz])}$

$L = \text{inductance (henry [H])};\ C = \text{capacitance (faraday } [F])$

For a pure resistor or circuit with a steady current, impedance is equal to the resistance: $Z = R$. Recall that Ohm's law is $I = (\frac{V}{R})$. A generalized form of Ohm's law is $V_{rms} = I_{rms}Z$, where the subscript *rms* means "root mean square" voltage or current for a time-average voltage/current in an alternating current (AC) system.

Consider a scenario in which an AC signal acts as a power source and is connected to one resistor and one capacitor. For a circuit with a frequency of 60 Hz and a 120 V_{rms} signal, if the capacitance (C) is 7 microfarad (μF), and the resistance is 300 Ω, what is the current in the circuit? Because there is no mention of an inductor, X_L is 0.

$$X_C = \frac{1}{2\pi f C} = \frac{1}{2\pi(60\text{ Hz})(7 \times 10^{-6}\text{ F})} = 378.9\ \Omega$$

$$Z = \sqrt{R^2 + (X_\text{L} - X_\text{C})^2} = \sqrt{(300\ \Omega)^2 + (-378.9\ \Omega)^2} = 483.3\ \Omega$$

$$I = \frac{V_{rms}}{Z} = \frac{120\text{ V}}{483.3\ \Omega} = 0.248\text{ A}$$

Energy

Energy may be defined as the capacity to do work. It can be transferred between objects, comes in a multitude of forms, and can be converted from one form to another. The **Law of Conservation of Energy** states, "Energy can neither be created nor be destroyed."

Electricity is a form of energy, like heat or movement, that can be harnessed to perform useful work. Electrical energy results from the electric force that exists between atoms and molecules with electrical charge, which is associated with the atomic structure of those substances. Atoms contain various subatomic particles. Protons, which are in the nucleus, carry a +1 charge. Electrons, which surround the outer part of an atom in orbitals or clouds, carry a –1 charge. Net charges of atoms or molecules occur when there is an imbalance in the number of electrons and protons. A net positive charge occurs when there are more protons than electrons, while a net negative charge results when there are more electrons than protons.

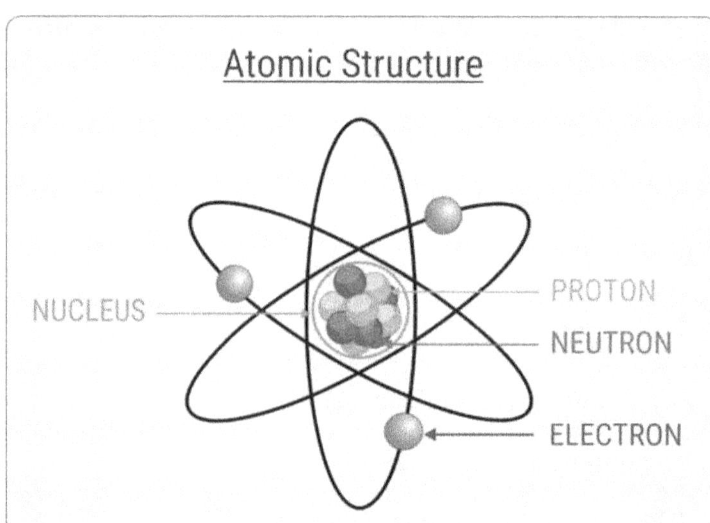

Atoms or molecules with electric charges that are the same experience a force that causes them to repel one another, while those with opposite charges attract each other. Therefore, two positive charges repel one another, two negative charges repel one another, but a positive and a negative charge attract one another. The unit of charge is denoted by C, the coulomb.

The Interaction of Charges

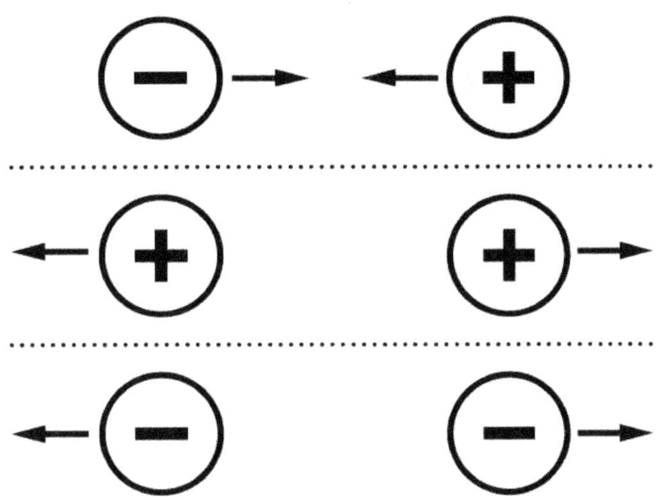

Similar to the Law of Conservation of Energy (which states that energy cannot be created or destroyed, only transferred from one form to another), there is a **conservation of charge** in the universe and in an isolated system. In a given isolated system, individual objects may experience a net loss or gain of charge with the transfer of charge from one object to another from within the system, but the overall charge within the system (or universe) cannot be created or destroyed. Individual positive or negative charges can be created or destroyed, but only in pairs (one positive with one negative), so that the net change in charge is zero.

Resistance

Electrical resistance, measured in ohms (Ω), is the amount of pressure inhibiting the flow of electrical current. Like friction, which slows the rate of movement, resistance dissipates energy and reduces the rate of flow or the movement of current. The amount of resistance that a given object contributes to a circuit depends on the properties of the object, particularly the material. Materials that are inherently more resistant inhibit the ease at which the electrons in the material's atoms can be displaced.

Some circuits have resistors built into them, which are specific electrical components designed to contribute a certain resistance to the circuit.

Resistance is inversely related to conductance, such that a highly conductive material has little resistance, and a material with high resistance has little conductance.

Materials vary in resistance because of the ease (or difficulty) with which electrons in the material's atoms can be displaced. The cross-sectional area and length of a given material also affect the resistance in a predictable relationship. The longer a given conductor is, the greater the resistance it provides; the greater the cross-sectional area (larger material), the less resistance there is. This relationship is quantitatively expressed as $R = \rho \cdot LA$, where ρ is the inherent resistivity of the specific conducting material, L is the length of the material, and A is the cross-sectional area.

Circuits

A **circuit** is a closed loop through which current can flow. A simple circuit contains a voltage source and a resistor. The current flows from the positive side of the voltage source through the resistor to the negative side of the voltage source. Note that if the switch is open or there is some other disconnected wire or break in continuity in the circuit, there will be no electromotive force; the circuit must be a closed loop to create a net flow of electrons from the voltage source through the wires and system.

Open and Closed Circuits

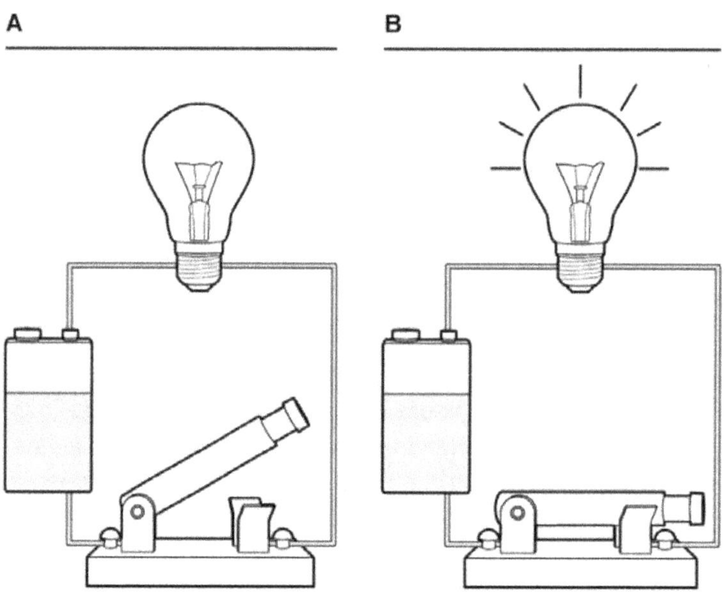

A: Open Circuit with no electric current

B: Closed Circuit with electric current flowing

An **electric circuit** is usually comprised of circuit elements joined by a wire or other object that allows an electric charge to move along the path without interruption—this moving electric charge is called an electric **current**. However, constant electric currents may only exist in a complete circuit if there is a voltage difference in the circuit. **Voltage** is literally the distance that a circuit's electrical force could move one electron, but voltage can be visualized as how much energy a certain part of the circuit has available to push around electrons. Electrons will flow from regions of higher voltage to regions of lower voltage, so it is the difference in voltages between two parts of a circuit that makes a current actually flow.

In other words, voltage difference is the difference in potential energy between two places measured in **volts (V)**, while a circuit is a closed path through which electrons can flow. Because every atom has positive charges that pull on electrons and resist their flow, most real circuits have a **resistance** level, measured in **ohms**, which is described as the opposition to the flow of electric charge. The amount of current that can flow through a circuit depends on the voltage difference and how well the wire resists the flow of electricity.

Principles of Radioactivity

Decay and Half-life

Radioactive decay is often described in terms of its **half-life ($t_{1/2}$)**, which is the time that it takes for half of the radioactive substance to react. Radioactive decay follows first-order kinetics and is similar to chemical reactions. The relationship between the half-life of a nuclide and its rate constant (k) is given by:

$$t_{1/2} = \frac{0.693}{k}$$

The larger the rate constant is for the nuclide, the shorter the half-life is.

For example, the radioisotope strontium-90 has a half-life of 28.8 years. If there are 10 grams of strontium-90 to start with, after 28.8 years, there would be 5 grams left. If you were only given the initial amount (10 g) and a final amount at time t (5 grams) and were asked to find the half-life at time t, then you would need to find the rate constant.

$$k = \frac{0.693}{t_{1/2}} = \frac{0.693}{28.8 \text{ years}} = 0.02406 \text{ years}^{-1}$$

Now solve for t using the following equation below:

$$\ln[\mathrm{B}]_t = -kt + \ln[\mathrm{B}]_0$$

$$\ln[\mathrm{B}]_t - \ln[\mathrm{B}]_0 = -kt$$

$$t = -\ln\left(\frac{[\mathrm{B}]_t}{[\mathrm{B}]_0}\right)\frac{1}{k}$$

The term $[B]_t$ is the amount at a time which happens to be the half-life because 5 g is 50% of the initial amount (10 g). $[\mathrm{B}]_0$ represents the initial amount at time zero. Therefore, $[\mathrm{B}]_0 = 10 \text{ g}$, so $[\mathrm{B}]_\mathrm{t} = 0.50(10 \text{ g}) = 5 \text{ g}$. The half-life is:

$$t = t_{1/2} = -\ln\left(\frac{5 \text{ g}}{10 \text{ g}}\right)\frac{1}{0.02406 \text{ years}^{-1}} = 28.8 \text{ years}$$

If asked to calculate the time given the half-life, where the final amount is half the initial amount, the answer will be the half-life (28.8 years). However, suppose you were asked to find the time at which strontium-90 decayed to 40% of its initial amount (4.0 g), then the time would be:

$$t = -\ln\left(\frac{0.40(\cancel{[\mathrm{B}]_0})}{\cancel{[\mathrm{B}]_0}}\right)\frac{1}{0.02406 \text{ days}^{-1}} = 38.1 \text{ days}$$

Note that 40% in decimal form is just $\frac{40}{100} = 0.40$ and that the final time is now longer than the half-life. It's also important to recognize how to rearrange the previous equation if asked to calculate the amount of substance at a time t, given the initial amount and the half-life:

$$[\mathrm{B}]_t = [\mathrm{B}]_0 e^{-kt}$$

For instance, how much of strontium-90 is left after $t = 38.1$ days if it has $t_{1/2} = 28.8$ days and an initial starting amount of 10 g? Solving for $[\mathrm{B}]_t$ gives:

$$[\mathrm{B}]_t = [\mathrm{B}]_0 e^{-kt} = (10\ \mathrm{g}) e^{-0.02406\ \mathrm{days}^{-1} \times 38.1\ \mathrm{days}} = 4.0\ \mathrm{g}$$

Note that half-life was given to solve for k, which was shown above.

Source Strength

Radioactivity is the emission of particles that result from the spontaneous breakdown of atomic nuclei. The general formulas are:

$$A = -\frac{dN}{dt}$$

$$A = \text{activity};\ N = \text{number of particles};\ t = \text{time}$$

Therefore, radioactivity is the volume of radiation emitted in a given amount of time. The strength of the radioactive source can vary and is important to know when considering hazardous, radioactive substances. The **source strength**, or activity, is the number of atoms that decay or disintegrate and emit radiation every second. The SI unit for source strength is the becquerel (Bq). The becquerel represents that amount of radioactive material, or breakdown of atomic nuclei, per second. The curie (Ci) represents the breakdown or decay of 3.7×10^{10} atoms (1 gram of radium [Ra] per second). One curie represents 37,037 megabecquerel (MBq). The general equation for source strength is:

$$Q = \frac{\lambda M N_A}{W}$$

$Q = \text{source strength (Bq)};\quad \lambda = \text{decay constant } (s^{-1});\quad M = \text{radionuclide mass (g)}$

$N_A = \text{Avogadro's number } 6.022 \times 10^{23}\ \frac{\text{atom}}{\text{gram-atom}};\quad W = \text{radionuclide atomic weight}$

If the half-life of radium-212 is 13.02 seconds, the rate of decay is:

$$\lambda = \frac{0.693}{13.02\ \mathrm{s}} = 0.05323\ \mathrm{s}^{-1}$$

The number 0.693, or the natural logarithm of 2, is the exponential rate of radioactive decay. The **atomic mass** refers to the mass of one atom, whereas **atomic weight** is the weighted average of all isotopes of atoms of the same element. There is one radioisotope, so the atomic mass of radium-212 is 212 atomic mass units (amu), which includes a specific number of protons and neutrons. Convert atomic mass units to grams using the conversion unit below.

$$M = 212\ \mathrm{amu} = \frac{1.66054 \times 10^{-24}\ \mathrm{g}}{1\ \mathrm{amu}} = 3.52 \times 10^{-22}\ \mathrm{g}$$

The atomic weight of radium is $W = 211.9998$ amu or 211.9998 grams/(gram-atom). The gram-atom is equivalent to a mole. Therefore, the source strength for the decay of radium-212 is:

$$Q = \frac{\lambda M N_A}{W} = \frac{(0.05324\ \text{s}^{-1})(3.52 \times 10^{-22}\ \text{g})(6.022 \times 10^{23}\ \text{atoms} \times (\text{g-atom})^{-1})}{211.9998\ \text{g} \times (\text{g-atom})^{-1}}$$

$Q = 0.0532$ Bq (source strength for one radium particle)

Concentration

Animal tests are often employed in laboratory studies to effectively predict **toxicity** in humans. Specimens such as rabbits and rodents are commonly used to test toxicity because they provide experimental reliability and are low cost and easily available. A toxicologist will attempt to create an experimental design that will mimic or approximate the potential exposure to humans. Some parameters include the dose levels, route of exposure, age of test animals, and sex of the animal. The **dose level** is the concentration of the substances received or exposed to over a period of time ($\text{Dose} = \text{concentration} \times \text{time}$). The **lethal dose ($LD_{50}$)** of a substance is the dose that is needed to kill 50%, or half, of the members of a population within a given time.

Acute toxicity tests provide information from a single exposure or multiple-dose exposures through oral/dermal administration over a brief period of time (<14 days). Sub-chronic toxicity tests **specify** if the toxicity came about through multiple exposures lasting from several weeks to months. Chronic toxicity tests determine toxicity for low exposure for a significant portion of the subject's life (12-24 months).

The Occupational Health and Safety Administration (OSHA) has defined the **permissible exposure limit (PEL)** as the legal limit for chemical substance exposure to an employee expressed in parts per million (ppm) and for physical agents, such as noise, in decibels (dB). The cumulative dose exposure for an 8-hour work shift (E) is:

$$\text{Dose } (D) = \frac{1}{8} \sum_{n}^{\text{exposure \#}} C_n T_n$$

D is the equivalent dose exposure for the working shift.

C is the concentration during a period of time, T, at a constant concentration.

T is the exposure time in hours for a specified concentration, C.

n is the number of exposure periods.

Chlorine concentrations near 5 ppm can result in respiratory issues, inflammation of the mucous membranes, and teeth erosion. OSHA attempted to revise the chlorine limit to 0.5 ppm within a 15-minute time period, as recommended by the National Institute for Occupational Safety and Health (NIOSH). The American Conference of Governmental Industrial Hygienists (ACGIH) has an 8-hour time-weighted average (TWA) limit of 0.5 ppm for chlorine. If an employee has the following exposures to chlorine, the acceptable exposure limit can be determined:

5 hours of exposure at 0.025 ppm

2 hours of exposure at 0.015 ppm

1 hour of exposure at 0.010 ppm

$$D = \frac{\sum_n^3 C_n T_n}{8} = \frac{1}{8}(C_1T_1 + C_2T_2 + C_3T_3) =$$

$$\frac{1}{8}\text{ hr }([0.025\text{ ppm}][5\text{ hr}] + [0.015\text{ ppm}][2\text{ hr}] + [0.010\text{ ppm}][1\text{ hr}]) = 0.021\text{ ppm}$$

Because 0.021 ppm is less than 0.5 ppm, the exposure limit is acceptable.

Inverse Square Law

Inverse square law is a rule of physics that shows a source's effect on another object considering the distance between the source and the other object. The greater the distance between the source and the other object, the lower the intensity that is placed on the other object. In other words, the farther away something is from the source, there is less intensity of the effects. The distance is inversely proportional to the dose.

Regarding radiation, the greater the distance between the radiation source and a person, the smaller the dose (or effect) is experienced by the person. Distance should be calculated. The calculated distance is the maximum distance from the source to properly influence the materials. Inverse square law has an equation that allows for determining the dose at a set distance.

$$I_2 = \frac{I_1(d_1)^2}{(d_2)^2}$$

I_1 is intensity at distance 1

I_2 is intensity at distance 2

$d_{1\text{ or }2}$ are distances at location 1 or 2

Order of operations is a mathematical principle that states the proper calculation sequence when completing a mathematical formula. Calculations are determined left to right in the sequence. This is also known as "PEMDAS": parentheses, exponents, multiplication and division (left to right), and addition and subtraction (left to right).

Examples:

1) A person is approximately 3 feet from a radiation source that measures 140 mrem/h. What is the dose rate if the person were 5 feet from the source?

$$I_2 = \frac{140\text{ mrem/h} \times (3\text{ ft})^2}{(5\text{ ft})^2}$$

$$I_2 = \frac{140\text{ mrem/h} \times 9\text{ ft}^2}{25\ ft^2}$$

$$I_2 = 50.4\text{ mrem/h}$$

If the person were 5 feet from the source, the exposure is 50.4mrem/h. The intensity decreased as the distance increased.

2) A person is approximately 5 feet from a radiation source that measures 170 mrem/h. What is the dose rate if the person were 8 feet from the source?

$$I_2 = \frac{170 \text{ mrem/h} \times (5 \text{ ft})^2}{(8 \text{ ft})^2}$$

$$I_2 = \frac{170 \text{ mrem/h} \times 25 \text{ ft}^2}{64 \text{ ft}^2}$$

$$I_2 = 66.41 \text{ mrem/h}$$

If the person were 8 feet from the source, the exposure is 66.41 mrem/h.

Storage Capacity Calculations

Load Capacity

Load capacity refers to the maximum stress that can be placed on a system under specific conditions for an extended time period. Load capacity is the system's capacity to perform an intended task when supporting a certain amount of weight (pounds) that can be safely placed on that system without causing it to collapse. OSHA has several rules that address the load capacity. Crane operators must be provided with information on the load capacity of a crane (29 CFR 1926). The load or storage capacity can also refer to the columns or beams used in the floor construction of a building. The load capacity is measured in pounds (lb) and is often labeled on operating equipment, tires, and beams. To determine the total load capacity, the load capacity for each component must be determined. In the analysis of steel columns, the load capacity is defined as:

$$P_{load} = F_a A; \quad F_a = \text{allowable compressive stress (ksi)};$$

$$A = \text{cross-sectional area of column (in}^2\text{)}$$

The kilopound per square inch (ksi) multiplied by the cross-sectional area will give the load capacity in units of kilopounds (klb). The applied stress is along the vertical, or y axis, of the column. If the actual load capacity (P_{actual}) is less than the allowable load capacity ($P_{allowable}$), the structural integrity of the column is acceptable. If $P_{actual} > P_{allowable}$, the column is overstressed. For one steel column with $F_a = 12.9$ ksi and $A = 12.0 \text{ in}^2$, the P_{load} is:

$$P_{load} = F_a A = 12.9 \frac{\text{klb}}{\text{in}^2} \times 12.0 \text{ in}^2 = 155 \text{ klb or } 155{,}000 \text{ lb-force}$$

The unit klb, or kip, is the kilopound. One kip is equal to 1000 pounds-force. If there are four steel columns supporting a platform, the total load capacity is:

$$\text{Total load capacity} = \textstyle\sum_n^{\#\text{ columns}} P_{load};\ n = \sum_n^4 P_{load};\ n = 155 \text{ kip } (4) = 620 \text{ kip}$$

Rigging and Load Calculations

Rigging and Slings

A crane or similar operating device is often rigged with a sling to lift a load. A **rigging hitch** connects the sling to the load. A **sling** is a rope, chain, or strap that connects the operating device to the load. A safety professional must ensure that the weight of the load does not exceed the safe **working load limit (WLL)**, or the maximum weight the sling is designed to handle. If the weight exceeds the maximum weight, the sling will break. The Mechanical Contractors Association of America (MCAA) defines three common types of hitches: vertical, basket, and choker. The vertical sling hitch is a sling that is tied to the top of the load; the angle between the sling and load is 90°. Other types of hitches, such as the bridle hitch, have been used to carry heavy loads. Several types of hitches are shown in the figure below.

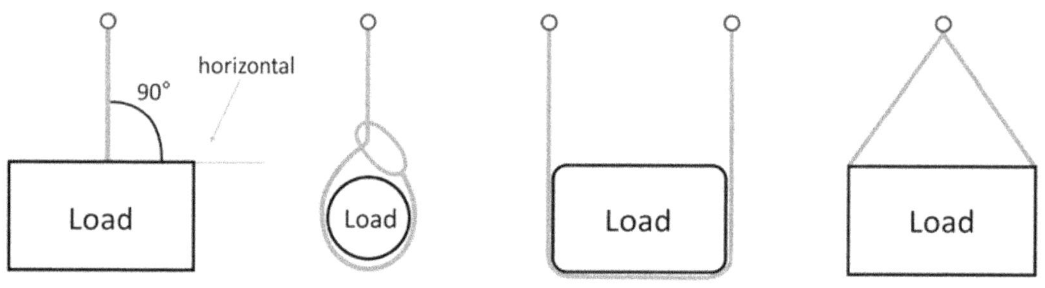

A vertical sling may have more than one sling. For instance, there may be two slings attached at each end of the load. The angle between each sling and the load is 90°. The sling is attached to a shackle that is connected to the load, and a hook connects the operating device to the sling. A choker hitch contains a single hitch or multiple hitches that wrap around the load. Each hitch contains an engineering loop at the end that wraps around the sling. A basket hitch loops or wraps the sling around the load; it allows the swing legs to act as two separate slings. The bridle hitch is a two-leg swing containing two legs/shackles at the end of each load linked with a cable. A hook is placed over the center of gravity. The sling tension, or tension factor, will vary depending on the number of legs and the angle of the sling to the load. The sling tension is defined as:

$$\text{Sling tension} = \frac{\text{load (kip)}}{\#\ \text{legs}\times\sin(\theta)}; \quad \theta = \text{angle between sling and load}$$

Ventilation and System Design

Ventilation involves the exchange of air from the outside of a building to the inside. Ventilation system design involves the sizing and selection of system components, such as the inlets, heaters, and fans. The two types of ventilation include mechanical and natural ventilation. Mechanical ventilation uses fans to provide airflow, whereas natural ventilation uses naturally occurring forces, such as wind. Mechanical ventilation consists of inlets or intakes, a split heater or cooler, a systems control, and an exhaust supply.

Heating, Ventilation, and Air Conditioning (HVAC)

Heating, ventilation, and air conditioning (HVAC) are part of the systems control that determines how much air is brought in from the outside. The ventilation component involves replacing or exchanging air inside a building, including controlling the air temperature, moisture, and oxygen replenishment. The ventilation system contains a particulate system that removes odors, airborne bacteria, dust/pollen, smoke, and gases, such as carbon dioxide.

The air change per hour (ACH), or air exchange rate, is a metric that indicates the number of times the HVAC completely fills a room with a specific volume of air. The volume is indicated below in the formula for ACH:

$$\text{Air change per hour } (ACH) = \frac{\text{air change}}{\text{time}} = \left(\frac{60 \text{ min}}{1 \text{ hr}}\right) \times \frac{Q \ (\text{ft}^3/\text{min})}{\text{area } (\text{ft}^2) \times \text{ceiling height (ft)}}$$

The term Q is the ventilation or volumetric flow rate with units of cubic feet per minute (CFM, or ft^3/min). This indicates the number of times all the air in the room is replaced with clean air over a period of time. The first term (60 min/hr) is the number of minutes per hour. The room volume, expressed in cubic feet (ft^3), is equal to the product of the area (ft^2) and the ceiling height (ft). If a purifier contains a 150 CFM airflow, what is the ACH for the purifier in a 200-square-foot room that has a ceiling height of 10 feet?

$$ACH = \frac{60 \text{ min}}{1 \text{ hr}} \times \frac{150 \ \text{ft}^3/\text{min}}{200 \text{ ft}^2 \times 10 \text{ ft}} = \frac{4.5}{1 \text{ hr}}$$

The HVAC system will completely fill a room with fresh air at a rate of 4.5 times per hour.

Noise Hazards

Noise is typically measured in decibels (dB), which is a logarithmic unit. The general equation for the sound intensity level (β) is:

$$\beta(\text{dB}) = 10 \times \log_{10}\left(\frac{I}{I_o}\right)$$

$$\beta(\text{dB}) = \text{sound intensity level}; \quad I = \text{sound level of the source } (\text{W/m}^2);$$

$$I_o = \text{base sound level at threshold of hearing} = 10^{-12} \text{ W/m}^2$$

The intensity of sound is measured in watts per square meter (W/m^2). A vacuum cleaner has a perceived loudness or sound level that is two times greater than the conversation between two people in an office, meaning the intensity levels will differ by 10 dB. For instance, if a vacuum cleaner has an intensity level of 70 dB, a conversation would measure 60 dB.

The permissible **occupational exposure limit (OEL)** set by OSHA is a TWA of 85 dB for an 8-hour workday. The hearing zone, defined by OSHA, is a 2-foot sphere surrounding the head. Above 85 dB, the worker must wear hearing protection. Above 100 dB, earmuffs and earplugs are required. Units of dBA refer to weighted measurements that have been adjusted to account for varying ear sensitivity. It has the same units as dB, but high and low frequencies have less weight compared to the standard Db scale. Regulatory nose limits are reported in dBA because they are correlated with hearing loss due to noise.

The noise dosimeter is a sound level meter or instrument that measures the noise exposure to a person's ears. The **noise dose**, measured by the dosimeter, can be used to find the 8-hour TWA for a specific period of time.

That equation is given by:

$$TWA = 16.61 \times \log_{10}\left(\frac{D}{100\%}\right) + 90$$

$$D = \text{dosimeter readout (\% dose);}$$

$$TWA = \text{8-hour TWA in decibels (dBA)}$$

A Noise Reduction Rating (NRR) provides a rating for hearing protection, which is based on the equipment's ability to decrease noise exposure. Earmuffs and plugs have an NRR between 25 and 30 dBA. The estimated exposure using hearing protection is:

$$\text{Estimated exposure (dBA)} = TWA\ \text{(dBA)} - (NRR - 7)$$

Suppose the TWA for a worker's environment is 110 dBA, and the worker is wearing earmuffs with an NRR equal to 25 dBA. The worker's estimated exposure is:

$$\text{Estimated exposure (dBA)} = 110 - (25 - 7) = 92\ \text{dBA}$$

Climate and Environmental Conditions

WetBulb Globe Temperature (WBGT) and Heat Stress

The National Weather Service (NWS) and National Oceanic and Atmospheric Administration (NOAA) define **WetBulb Globe Temperature (WBGT)** as a heat stress measurement, in degrees Fahrenheit, under direct sunlight. **Heat stress** involves a number of conditions whereby the body is under stress from heat exposure. For a worker, heat stress is the combination of metabolic heat, environmental heat factors, and the body's heat storage due to clothing. For environmental factors, the WBGT accounts for various conditions, such as wind speed, temperature, and humidity. Other factors, such as the sun angle and the solar radiation, or cloud cover, are included in the WBGT. The heat index accounts for temperature and humidity but is based on shaded regions. For workers who work or exercise under direct sunlight, OSHA, schools, and various agencies use the WGBT as a reference on handling workload in the sun. The following table provides a list of suggested actions and prevention methods.

WGBT (°F)	Effects when working/exercising in direct sunlight: The body will stress after a specified amount of time (minutes).	Preventive measures: When working/exercising in sunlight, the worker will take a break (in minutes) per hour.
<80	No precautions needed	No precautions needed
80-85	Stressed at 45	15-minute break
85-88	30	30
88-90	20	40
>90	15	45

For a WGBT greater than 90, a worker will be stressed in 15 minutes; the worker must take a 45-minute break. The WBGT is calculated using a special instrument that outputs a value in degrees Fahrenheit.

The WGBT gives the ratio of work to rest that workers should follow to avoid heat stress. The WGBT is given by the following calculation:

$$WBGT\ (°\mathrm{F}) = 0.7\ T_w + 0.2\ T_g + 0.1\ T_d$$

$$T_w = \text{wet-bulb temperaure (°F or °C)}$$

$$T_g = \text{global thermometer temperature (°F or °C)}$$

$$T_d = \text{actual air or dry-bulb temperature (°F or °C)}$$

The global thermometer temperature (T_g) is measured using a thermometer placed within a black globe that is designed to exclude light effects. The thermometer is used to calculate solar radiation. The wet-bulb temperature (T_w) is a complex formula that accounts for a cooling effect when water evaporates in the air. The dry-bulb temperature (T_d) is the temperature of the air that can be taken with a simple thermometer.

Wind Chill

Wind chill is the approximate temperature experienced by a bare body (generally a body part) based on the combination of wind speed and temperature. It is the number of calories lost in one hour on a square meter of a surface kept at 91.4°F. Wind chill may be calculated by using the following formula:

$$Wind\ Chill\ (°F) = 35.74 + 0.6215T - 35.75(V^{0.16}) + 0.4275T(V^{0.16})$$

T is air temperature (°F)

V is wind speed (mph)

Example:

1) Determine the wind chill for conditions with a 25°F temperature and 20mph wind speed.

$$\text{Wind Chill} = 35.74 + 0.6215(25) - 35.75(20^{0.16}) + 0.4275(25)(20^{0.16})$$

$$35.74 + 15.54 - 35.75(1.61) + 0.4275(25)(1.61)$$

$$35.74 + 15.54 - 57.56 + 17.21$$

$$10.84 \rightarrow 11$$

$$\text{Wind Chill} = 11$$

Wind Chill Chart

Temperature (°F)

Wind(mph) / Calm	40	35	30	25	20	15	10	5	0	-5	-10	-15	-20	-25	-30	-35	-40	-45
5	36	31	25	19	13	7	1	-5	-11	-16	-22	-28	-34	-40	-46	-52	-57	-63
10	34	27	21	15	9	3	-4	-10	-16	-22	-28	-35	-41	-47	-53	-59	-66	-72
15	32	25	19	13	6	0	-7	-13	-19	-26	-32	-39	-45	-51	-58	-64	-71	-77
20	30	24	17	11	4	-2	-9	-15	-22	-29	-35	-42	-48	-55	-61	-68	-74	-81
25	29	23	16	9	3	-4	-11	-17	-24	-31	-37	-44	-51	-58	-64	-71	-78	-84
30	28	22	15	8	1	-5	-12	-19	-26	-33	-39	-46	-53	-60	-67	-73	-80	-87
35	28	21	14	7	0	-7	-14	-21	-27	-34	-41	-48	-55	-62	-69	-76	-82	-89
40	27	20	13	6	-1	-8	-15	-22	-29	-36	-43	-50	-57	-64	-71	-78	-84	-91
45	26	19	12	5	-2	-9	-16	-23	-30	-37	-44	-51	-58	-65	-72	-79	-86	-93
50	26	19	12	4	-3	-10	-17	-24	-31	-38	-45	-52	-60	-67	-74	-81	-88	-95
55	25	18	11	4	-3	-11	-18	-25	-32	-39	-46	-54	-61	-68	-75	-82	-89	-97
60	25	17	10	3	-4	-11	-19	-26	-33	-40	-48	-55	-62	-69	-76	84	-91	-98

Fall Protection Calculations

Fall protection is the measure or measures taken to prevent individuals from falling. OSHA dictates fall protection is applicable to work done above four feet (general industry), five feet (marine), six feet (construction), and specifically, ten feet for scaffolding.

Personal Fall Arrest System (PFAS)

PFAS setups require a lot of planning and testing to ensure they will work as intended. In contrast with guardrail systems, a PFAS is much more complex in every regard and should be used only when guardrails are not a feasible means of fall protection (for example, on sloped working surfaces such as pitched roofs).

A PFAS includes a full-body harness (body belts are not acceptable), lanyard, other devices and connectors, such as D-rings, and the anchorage point itself. All of these components must meet strength requirements and be tested to ensure safe function. Five thousand is a good number to remember with PFAS.

An anchorage must support 5000 pounds per worker attached or a safety factor of two. Since the latter involves additional testing with a representative weight and desired drop distance to ensure it will be effective in a given circumstance, most fall protection specialists will select an anchorage point verified to support 5000 pounds per worker. Note that 5000 pounds is not referring to total weight but the weight in force that will be transmitted to the anchorage point when the fall is arrested.

Lanyards and vertical lifelines must have at least 5000 pounds of breaking strength. D-rings and snap hooks must have at least 5000 pounds of tensile strength.

An ASP must understand that, even when functioning properly, PFAS can still injure an employee when the fall is arrested, and a fall restraint system should be used whenever practical to do so.

To minimize the likelihood of a fall arrest causing an injury, OSHA permits a drop of no more than six feet. The maximum fall distance allowed is based on the fall distance itself, not the length of the lanyard alone, so the selection of the anchorage point and all components must be chosen carefully to ensure that an employee cannot fall more than six feet. Shock-absorbing lanyards will reduce the likelihood of an injury during a fall arrest because they reduce the force transmitted to the employee's body. Shock-absorbing units destroy themselves when an arrest occurs by tearing (stretching.) Since they will lengthen when arresting a fall, this must be taken into consideration when planning to keep the maximum fall distance to six feet. Shock-absorbing lanyards can never be used after they have been stretched in a fall arrest situation.

Stairs and Ladders

Work activities sometimes require the use of stairs and ladders; it may be necessary to have stairs built for the crew to use during the building project. OSHA guidelines (1926.1052(a)(2)) state that stair construction requires a uniform riser height and a tread depth of no more than a ¼ inch variation of each element. The stair slope or angle must fall between 30 and 50 degrees. Stairs with the presence of four or more risers (or exceeding 30 inches in height) require at least one handrail that's a minimum of 36 inches in height above each stair nosing. The handrail must be capable of withstanding at least 200 pounds of side force, and a midrail must be present halfway between the handrail and the steps.

Ladders should be set using a 4:1 ratio where the point of contact at height distance is four times greater than the point of contact distance at the base (away from the wall). When in position and secure, ladders must extend three feet past the surface of the upper landing. Safe use requires the worker to face the ladder at all times, use at least one hand to grasp the ladder (three points of contact), avoid carrying any load that would cause loss of balance, and never use or perform work on the top rung.

OSHA requires fall protection at a height at above 24 feet for fixed ladders. Workers who climb less than 24 feet in height on a fixed ladder that still extends up to at least 24 feet also need to have cages, ladder safety devices, wells, or self-retracting lifelines provided. Extension ladders less than 36 feet should maintain a minimum overlap of three feet; those between 36 and 48 feet should maintain a minimum overlap of four feet. Ladders between 48 and 60 feet require a minimum overlap of five feet. A competent person should inspect step ladders and extension ladders before each use. When a ladder is removed from service, subsequent repairs should be done to return the ladder to its original design criteria.

General Physics Concepts

Force

Isaac Newton's three laws of motion describe how the acceleration of an object is related to its mass and the forces acting on it. The three laws are:

- Unless acted on by a force, a body at rest tends to remain at rest; a body in motion tends to remain in motion with a constant velocity and direction.

- A force that acts on a body accelerates it in the direction of the force. The larger the force, the greater the acceleration; the larger the mass, the greater its inertia (resistance to movement and acceleration).
- Every force acting on a body is resisted by an equal and opposite force.

To understand Newton's laws, it's necessary to understand forces. These forces can push or pull on a mass, and they have a magnitude and a direction. Forces are represented by a vector, which is the arrow lined up along the direction of the force with its tip at the point of application. The magnitude of the force is represented by the length of the vector.

The figure below shows a mass acted on or "pushed" by two equal forces (shown here by vectors of the same length). Both vectors "push" along the same line through the center of the mass, but in opposite directions. What happens?

A Mass Acted on by Equal and Opposite Forces

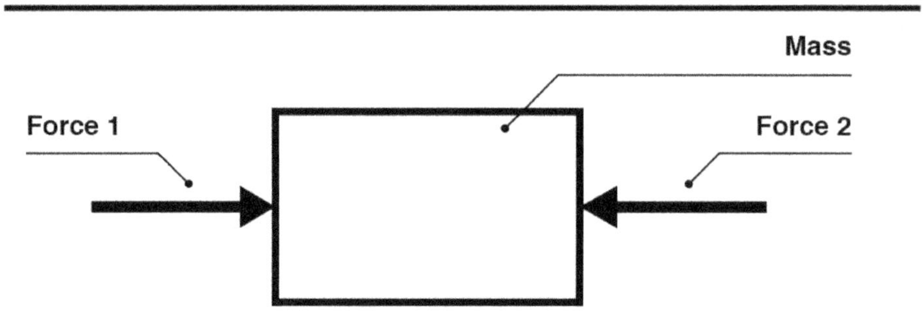

According to Newton's third law, every force on a body is resisted by an equal and opposite force. In the figure above, Force 1 acts on the left side of the mass. The mass pushes back. Force 2 acts on the right side, and the mass pushes back against this force too. The net force on the mass is zero, so according to Newton's first law, there's no change in the **momentum** (the mass times its velocity) of the mass. Therefore, if the mass is at rest before the forces are applied, it remains at rest. If the mass is in motion with a constant velocity, its momentum doesn't change. So, what happens when the net force on the mass isn't zero, as shown in the figure below?

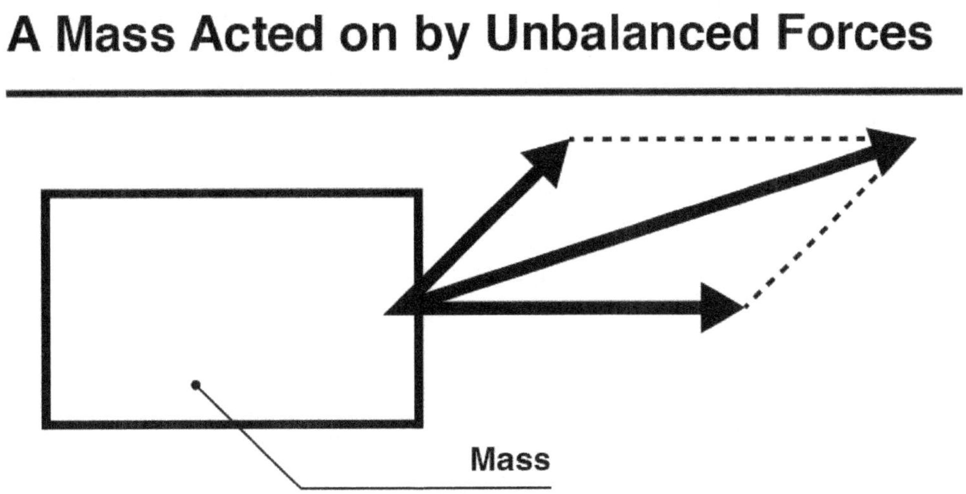

Notice that the forces are vector quantities and are added geometrically the same way that velocity vectors are manipulated.

Here in the figure above, the mass is pulled by two forces acting to the right, so the mass accelerates in the direction of the net force. This is described by Newton's second law:

$$\text{Force} = \text{Mass} \times \text{Acceleration}$$

The force (measured in *newtons*) is equal to the product of the mass (measured in kilograms) and its acceleration (measured in meters per second squared or meters per second, per second). A better way to look at the equation is dividing through by the mass:

$$\text{Acceleration} = \frac{\text{Force}}{\text{Mass}}$$

This form of the equation makes it easier to see that the acceleration of an object varies directly with the net force applied and inversely with the mass. Thus, as the mass increases, the acceleration is reduced for a given force. To better understand, think of how a baseball accelerates when hit by a bat. Now imagine hitting a cannonball with the same bat and the same force. The cannonball is more massive than the baseball, so it won't accelerate very much when hit by the bat.

In addition to forces acting on a body by touching it, gravity acts as a force at a distance and causes all bodies in the universe to attract each other. The **force of gravity (F_g)** is proportional to the masses of the two objects (m and M) and inversely proportional to the square of the distance (r^2) between them (and G is the proportionality constant).

This is shown in the following equation:

$$F_g = G\frac{mM}{r^2}$$

The force of gravity is what causes an object to fall to Earth when dropped from an airplane. Understanding gravity helps explain the difference between mass and weight. Mass is a property of an object that remains the same while it's intact, no matter where it's located. A 10-kilogram cannonball has the same mass on Earth as it does on the moon. On Earth, it weighs 98.1 newtons because of the attractive force of gravity, so it accelerates at 9.81 m/s^2. However, on the moon, the same cannonball has a weight of only about 16 newtons. This is because the gravitational attraction on the moon is approximately one-sixth that on Earth. Although Earth still attracts the body on the moon, it's so far away that its force is negligible.

Another way to understand Newton's second law is to think of it as an object's change in momentum, which is defined as the product of the object's mass and its velocity:

$$\text{Momentum} = \text{Mass} \times \text{Velocity}$$

Acceleration and Velocity

While an object's speed measures how fast the object's position will change in a certain amount of time, an object's **acceleration** measures how fast the object's speed will change in a certain amount of time. Acceleration can be thought of as the change in velocity or speed (Δv) divided by the change in the time (Δt).

$$\text{acceleration } (a) = \frac{\Delta v}{\Delta t}$$

Velocity is measured in meters/second and time is measured in seconds, so the standard unit for acceleration is meters / second2 (m/s2).

$$\frac{\text{meters}/\text{second}}{\text{second}} = \frac{\text{meters}}{\text{second}^2} = \frac{\text{m}}{\text{s}^2}$$

Acceleration is expressed by using both magnitude and direction, so it is a vector quantity like velocity. Acceleration is present when an object is slowing down, speeding up, or changing direction, since these represent instances where velocity is changing. This means that forces like friction and gravity accelerate objects and increase or decrease their velocities over time.

Momentum

Motion creates something called **momentum**. This is a calculation of an object's mass multiplied by its velocity. Momentum can be described as the amount an object wants to continue moving along its current course. Momentum in a straight line is called linear momentum. Just as energy can be transferred and conserved, so too can momentum.

Changing the expression of Newton's second law of motion yields a new expression.

$$\text{Force } (F) = ma = m \times \frac{\Delta v}{\Delta t}$$

If both sides of the expression are multiplied by the change in time, the law produces the impulse equation.

$$F\Delta t = m\Delta v$$

This equation shows that the amount of force during a length of time creates an **impulse**. This means that if a force acts on an object during a given amount of time, it will have a determined impulse. However, if the same change in velocity happens over a longer amount of time, the required force is much smaller, due to the conservation of momentum.

$$p = mv$$

Linear momentum, p, is found by multiplying the mass of an object by its velocity. Since momentum, like mass and energy, is conserved, Newton's 2nd law can be restated for multiple objects. In this form, it can be used to understand the energy of objects that have interacted, since the conservation of momentum implies that the momentum before and after an interaction must be the same. This is best demonstrated in the case of elastic collision, where an object of mass m_1 with velocity v_1 collides with an object of mass m_2 with velocity v_2 and both object end with velocities v_1' and v_2', respectively.

$$m_1v_1' + m_2v_2' = m_1v_1 + m_2v_2$$

Friction

Friction is a force that opposes motion. It can be caused by several materials; there are even frictions caused by air or water. Whenever two differing materials touch, rub, or pass by each other, this can create friction, or an oppositional force. To move an object across a floor, the force exerted on the object must overcome the frictional force keeping the object in place. Friction is also why people can walk on surfaces. Without the oppositional force of friction to a shoe pressing on the floor, a person would not be able to grip the floor to walk—similar to the challenges of walking on ice. Without friction, shoes slip and are unable to help people push forward and walk. Start-up and sliding friction forces are calculated in the same way: normal force (or weight) times the friction coefficient.

Pushing a block horizontally along a rough surface requires work. In this example, the work needs to overcome the force of friction, which opposes the direction of the motion and equals the weight of the block times a **friction factor (f)**. The friction factor is greater for rough surfaces than smooth surfaces, and it's usually greater *before* the motion starts than after it has begun to slide. These terms are illustrated in the figure below.

Pushing a Block Horizontally Against the Force of Friction

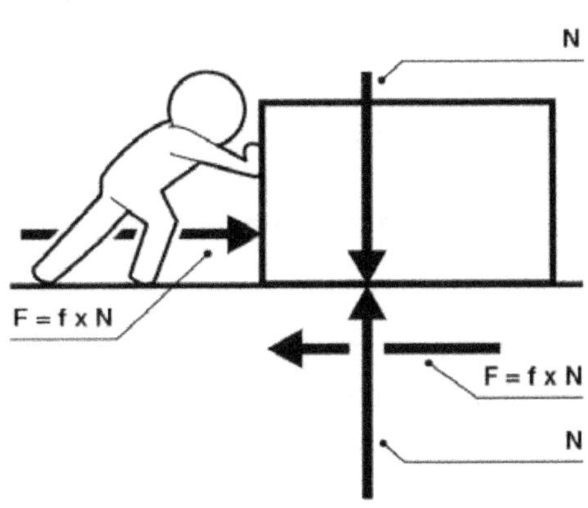

When pushing a block, there's no increase in potential energy since the block's elevation doesn't change. Expending the energy to overcome friction is "wasted" in the generation of heat. Yet, to move a block from point A to point B, an energy cost must be paid. However, friction isn't always a hindrance. In fact, it's the force that makes the motion of a wheel possible.

Financial Principles

Cost-Benefit Analysis

A **cost-benefit analysis (CBA)** is an important factor in many business decisions. A CBA compares the cost of a particular option with the benefits it will bring to the organization. A CBA has two main uses. First, it helps to determine whether a particular option is worthwhile (Do the benefits sufficiently offset the costs?). Secondly, it's a method of comparison when making a decision that has several options.

Of course, any cost-benefit analysis involves a certain level of uncertainty because it's predicting future values under future conditions. For example, a change in the cost of a certain resource or the exchange rate of foreign currency could impact the results of a CBA. For this reason, a CBA usually includes a sensitivity analysis, which determines how much a change in uncertain variables will affect the CBA. This sensitivity analysis takes into account the expected conditions (what will happen if everything proceeds according to the status quo?) as well as worst-case conditions (what will happen if all possible problems arise in this situation?). In this way, a CBA can also reveal the level of risk involved in a decision. An option that appears attractive at first may seem less certain after a sensitivity analysis.

A cost-benefit analysis can also be approached differently depending on the view of the analysis—short-, mid-, or long-term. For many business decisions, the costs are upfront while the benefits may appear immediately or after a longer period of time. For this reason, a short-term and long-term CBA could yield very different results. If an organization needs a quick return on benefits, it might place more emphasis on a short-term CBA. However, if it's willing to wait longer to reap the benefits of a decision, it might compare its options based on long-term CBAs.

Cost of Risk

Cost of risk refers to the management of financial risk while incurring losses due to financial risk. Expenses associated with financial management of risk include risk losses, insurance premiums, guaranteed financials, internal expenses, taxes, and other likely financial obligations.

Generally, overall cost of risk includes three categories: (1) preventative costs; (2) direct costs; and (3) indirect costs. The three categories aggregated provide the cost of risk. Many believe that insurance alone covers the cost of risk, but that is a misconception. Research has shown that costs used in preventing claims or losses has reduced the direct costs to the organization. Generally, the preventative costs represent a lower amount than the direct costs over the long term of the organization's operation. It is beneficial to implement structured preventative measures to improve the financial stability of an organization.

Preventative costs refer to the expenses of preventative measures implemented to reduce an organization's risk. This includes such things as safety equipment, employee wellness programs, employee training, and safety management implemented by the organization.

Direct costs refer to the expenses that are directly attributable to risk and have developed into a harm or loss. This includes such things as insurance, administrative fines, legal expenses, and productivity loss.

Indirect costs refer to the expenses that are not directly tied to risk but have an effect on risk costs. This includes such things as goodwill, reputation with external entities, and decreased morale within the organization.

Life Cycle Cost

The total cost of a specific asset, such as a building or device, can change over a period of time. **Life cycle costing (LCC)** is an economic analysis method that involves the selection of alternative materials that will impact pending and future costs. LCC will compare initial investment options and identify the least cost alternatives for a time period spanning 20 years. LCC has application to building design energy conservation. The National Institute of Standards and Technology (NIST) has prepared LCC-guided manuals. Several practices are listed for building design and include defining alternatives for LCC. The number of costs will depend on the type of asset. In general, for a particular asset, the costs include the initial cost (C_i), installation cost (C_{inst}), energy costs (C_e), operational costs (C_o), maintenance costs (C_m), downtime costs (C_t), and disposal costs (C_d). The LCC formula is given by:

$$LCC = C_i + C_{inst} + C_e + C_o + C_m + C_t + C_d$$

Suppose a company wants to purchase a bulb that will last 10,000 hours. The company has the option of buying a $23 100-watt incandescent light bulb with a life expectancy of 1000 hours or a $100 25-watt fluorescent bulb with a life expectancy of 10,000 hours. The economist must calculate the LCC for both alternatives to find the more cost-effective bulb. For simplicity, consider the initial and operational costs to use a light bulb. For a light bulb, the initial (*Ci*) and operational (*Co*) cost in dollars can be calculated using the following equations.

$$Ci = \text{cost of asset} \times \left(\frac{\text{usage hours}}{\text{life expectancy}}\right) = \$23 \times \left(\frac{10{,}000\text{ hr}}{1000\text{ hr}}\right) = \$230$$

A 100-watt light bulb will consume 100 watts per hour. In the United States, the average cost of 1 kilowatt per hour (kWh) is $0.12. The number of hours to use 1 kWh is equal to 1000 watts divided by 100 watts/hour, which is equal to 10 hours. The operational costs are:

$$Co = \text{cost of } 1\ \frac{\text{kWh}}{\text{hour}} \times \left(\frac{\text{usage hours}}{\text{hours to use 1 kWh}}\right) = \$0.12 \times \left(\frac{10{,}000\text{ hr}}{\frac{1000}{100}\text{ hr}}\right) = \$120$$

$$LCC_{bulb} = Ci + Co = \$230 + \$120 = \$350$$

Return on Investment

Another important metric is **return on investment (ROI)**. ROI is generally expressed as a ratio or percentage comparing the gains of a particular investment with its initial investment price. In other words, ROI measures the ratio between an investment's profit and its cost. ROI is particularly useful in helping an organization evaluate the overall value of a given investment. For example, one investment may yield a high return, but perhaps the initial investment is costly as well. Another investment with a much lower yield also has a far lower initial cost—so the cheaper investment might actually have a higher ROI than the high-return investment. This metric can help an organization to devote its financial resources to investments with the highest ROI.

Effects of Losses

The economic impact due to work-related injury costs and time lost is significant and amounts to billions of dollars. A 2013 study by the US Bureau of Labor Statistics and the National Academy of Social Insurance indicated that the annual cost US businesses paid due to workplace injuries was $62 billion. The study referred to workers who missed six or more days of work due to an injury. Overexertion was the leading cause of disabling work injuries, which accounted for 25% of the injuries.

Studies by the National Safety Council (NSC) have measured work injury costs that impact society, which is comparable to other economic measures, such as gross domestic product or per capita income. In 2019, the NSC estimated that the cost of work injuries was $171 billion, which includes employers' uninsured costs of $13.3 billion, medical expenses of $35.5 billion, administrative expenses of $59.7 billion, and wage and productivity losses of $53.9 billion. The employer's uninsured costs include the time value lost by workers and the cost to investigate accidents and write accident reports. Work-related losses due to workers injured in vehicle wrecks were $5 billion, and fire-related incidents amounted to $3.7 billion. For a per-worker cost, the 2019 NSC study estimated that medical costs due to injury were $42,000, and a cost per death was $1.22 million. The 2019 study indicated that the lost time in days due to injuries was equal to $70 million, which excludes time lost when the worker was injured as well as the time needed for medical treatment and checkups.

Descriptive Statistics

Descriptive statistics offer an understanding of properties of a data set. Descriptive statistics involves analyzing a collection of data to describe its broad properties such average (or mean), what percent of the data falls within a given range, and other such properties. An example of this would be taking all of the test scores from a given class and calculating the average test score. Descriptive statistics entails examining the center, spread, and shape of the sample data.

Central Tendency

The center of a set of data (statistical values) can be represented by its mean, median, or mode. These are sometimes referred to as measures of central tendency.

Mean

The first property that can be defined for this set of data is the **mean**. This is the same as the average To find the mean, add up all the data points, then divide by the total number of data points. For example, suppose that in a class of 10 students, the scores on a test were 50, 60, 65, 65, 75, 80, 85, 85, 90, 100. Therefore, the average test score will be:

$$\frac{50+60+65+65+75+80+85+85+90+100}{10}=75.5$$

The mean is a useful number if the distribution of data is normal (more on this later), which roughly means that the frequency of different outcomes has a single peak and is roughly equally distributed on both sides of that peak. However, it is less useful in some cases where the data might be split or where there are some **outliers**. Outliers are data points that are far from the rest of the data. For example, suppose there are 10 executives and 90 employees at a company. The executives make \$1000 per hour, and the employees make \$10 per hour.

Therefore, the average pay rate will be:

$$\frac{\$1000 \times 11 + \$10 \times 90}{100} = \$119 \text{ per hour}$$

In this case, this average is not very descriptive since it's not close to the actual pay of the executives *or* the employees.

Median

Another useful measurement is the **median**. In a data set, the median is the point in the middle. The middle refers to the point where half the data comes before it and half comes after, when the data is recorded in numerical order. For instance, these are the speeds of the fastball of a pitcher during the last inning that he pitched (in order from least to greatest):

90, 92, 93, 93, 95, 96, 97, 97, 97

There are nine total numbers, so the middle or **median** number is the 5th one, which is 95.

In cases where the number of data points is an even number, then the average of the two middle points is taken. In the previous example of test scores, the two middle points are 75 and 80. Since there is no single point, the average of these two scores needs to be found. The average is:

$$\frac{75 + 80}{2} = 77.5$$

The median is generally a good value to use if there are a few outliers in the data. It prevents those outliers from affecting the "middle" value as much as when using the mean.

Since an outlier is a data point that is far from most of the other data points in a data set, this means an outlier also is any point that is far from the median of the data set. The outliers can have a substantial effect on the mean of a data set, but they usually do not change the median or mode, or do not change them by a large quantity. For example, consider the data set (3, 5, 6, 6, 6, 8). This has a median of 6 and a mode of 6, with a mean of $\frac{34}{6} \approx 5.67$. Now, suppose a new data point of 1000 is added so that the data set is now (3, 5, 6, 6, 6, 8, 1000). The median and mode, which are both still 6, remain unchanged. However, the average is now $\frac{1034}{7}$, which is approximately 147.7. In this case, the median and mode will be better descriptions for most of the data points.

Outliers in a given data set are sometimes the result of an error by the experimenter, but oftentimes, they are perfectly valid data points that must be taken into consideration.

Mode

One additional measure to define for X is the **mode**. This is the data point that appears most frequently. If two or more data points all tie for the most frequent appearance, then each of them is considered a mode. In the case of the test scores, where the numbers were 50, 60, 65, 65, 75, 80, 85, 85, 90, 100, there are two modes: 65 and 85.

Variability

Spread

Methods for determining the spread of the sample include calculating the range and standard deviation for the data. The **range** is calculated by subtracting the lowest value from the highest value in the set. The **standard deviation** of the sample can be calculated using the formula:

$$= \sqrt{\frac{\Sigma(x - \bar{x})^2}{n - 1}}$$

$\bar{x}$ = sample mean
n = sample size

Shape

The shape of the sample when displayed as a histogram or frequency distribution plot helps determine if the sample is normally distributed, symmetrical, or skewed (asymmetrical); the shape can also indicate the sample's level of kurtosis. **Kurtosis** is a measure of whether the data are heavy-tailed (high number of outliers) or light-tailed (low number of outliers).

Normal Distributions

Normal distribution refers to variable data in which most of the data points are near the mean. Approximately 68% of the data values are within one standard deviation of the mean, 95% within two standard deviations, and 99.7% within three standard deviations. This is referred to as the 3-sigma rule. Since normal distributions are shaped like bells, this curve is known as a bell curve. The probability density function for a normal distribution is:

$$P(x) = \frac{e^{\frac{-[\![(x-\mu)^2]\!]}{2\sigma^2}}}{\sigma\sqrt{2\pi}}$$

μ = mean
σ = standard deviation

The standard normal distribution has a mean of 0, a standard deviation of 1, and the total area under its curve is 1.

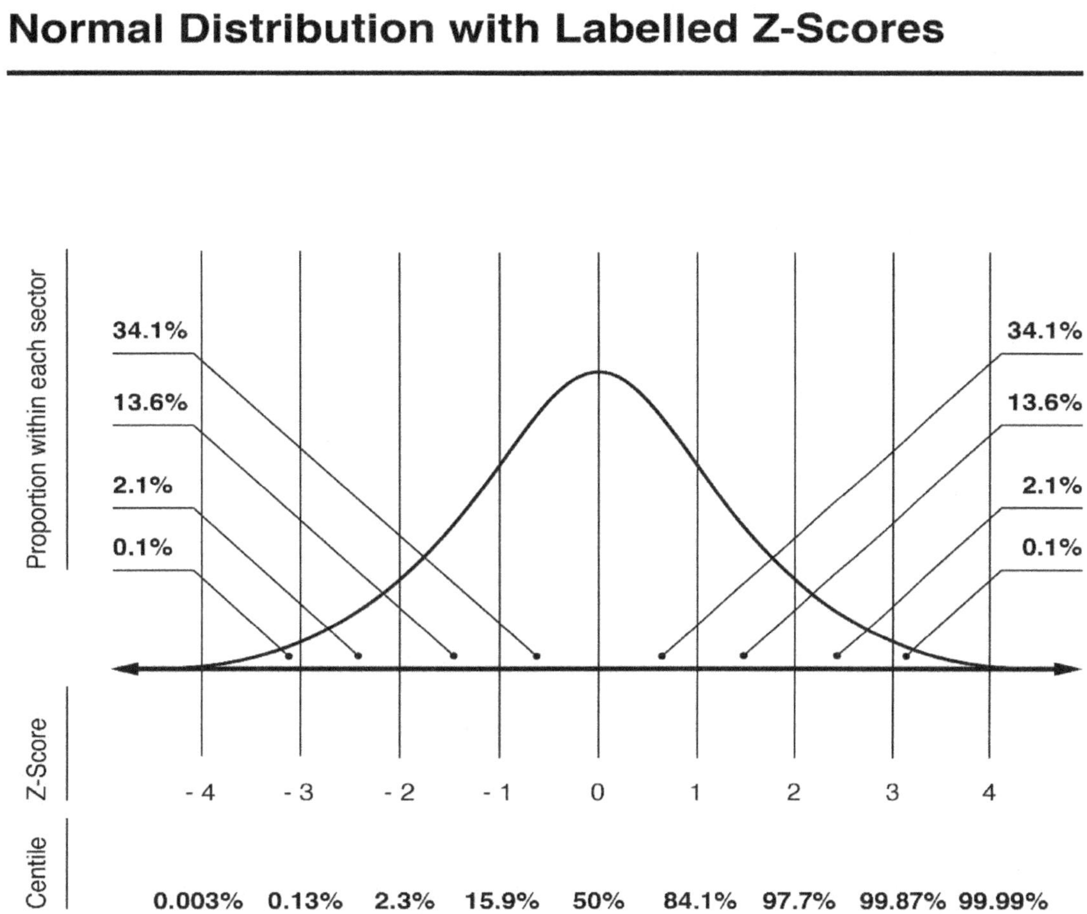

Z-scores for a point in a normal distribution measure the number of standard deviations the value is from the mean. They are used to find the area between two points on a curve. A z-score can be calculated using the formula: $z = \frac{(x-\mu)}{\sigma}$. Comparing the total area under the curve to the area between two points on the curve can determine the probability that a random data point lies between these two points.

Normal probability plots identify if a sample data set approximates a normal distribution. Normal distribution is a base assumption for certain statistical analyses and inferences. Therefore, testing for normality is important. The axes of the normal probability plot consist of the sample data and normal order statistic medians. If the data models normality, the points will form a linear pattern. The closer the points are to forming a straight line, the closer the data set is to a normal distribution.

Probability

Probability is the chance or extent to which something is likely to happen. Generally, the **probability** of a particular event happening is between 0 and 1.

Given a set of possible outcomes X, a **probability distribution** on X is a function that assigns a probability to each possible outcome. If the outcomes are $(x_1, x_2, x_3, \dots x_n)$, and the probability distribution is p, then the following rules are applied.

- $0 \leq p(x_i) \leq 1$, for any i.
- $\sum_{i=1}^{n} p(x_i) = 1$.

In other words, the probability of a given outcome must be between zero and 1, while the total probability must be 1.

If $p(x_i)$ is constant, then this is called a **uniform probability distribution**, and $p(x_i) = \frac{1}{n}$. For example, on a six-sided die, the probability of each of the six outcomes will be $\frac{1}{6}$.

If seeking the probability of an outcome occurring in some specific range A of possible outcomes, written $P(A)$, add up the probabilities for each outcome in that range. For example, consider a six-sided die, and figure the probability of getting a 3 or lower when it is rolled. The possible rolls are 1, 2, 3, 4, 5, and 6. So, to get a 3 or lower, a roll of 1, 2, or 3 must be completed. The probabilities of each of these is $\frac{1}{6}$, so add these to get:

$$p(1) + p(2) + p(3) = \frac{1}{6} + \frac{1}{6} + \frac{1}{6} = \frac{1}{2}$$

An outcome occasionally lies within some range of possibilities B, and the probability that the outcomes also lie within some set of possibilities A needs to be figured. This is called a conditional probability. It is written as $P(A|B)$, which is read "the probability of A given B." The general formula for computing conditional probabilities is:

$$P(A|B) = \frac{P(A \cap B)}{P(B)}$$

However, when dealing with uniform probability distributions, simplify this a bit. Write $|A|$ to indicate the number of outcomes in A. Then, for uniform probability distributions, write:

$$P(A|B) = \frac{|A \cap B|}{|B|}$$

Recall that $A \cap B$ means "A intersect B" and consists of all of the outcomes that lie in both A and B. This means that all possible outcomes do not need to be known. To see why this formula works, suppose that the set of outcomes X is $(x_1, x_2, x_3, \dots x_n)$, so that $|X| = n$. Then, for a uniform probability distribution:

$$P(A) = \frac{|A|}{n}$$

However, this means:

$$(A|B) = \frac{P(A \cap B)}{P(B)} = \frac{\frac{|A \cap B|}{n}}{\frac{|B|}{n}} = \frac{|A \cap B|}{|B|}$$

Note that the *n*'s cancel out.

For example, suppose a die is rolled, and it is known that it will land between 1 and 4. However, how many sides the die has is unknown. Figure the probability that the die is rolled higher than 2. To figure this, $P(3)$ or $P(4)$ does not need to be determined, or any of the other probabilities, since it is known that a fair die has a uniform probability distribution. Therefore, apply the formula $\frac{|A \cap B|}{|B|}$. So, in this case B is (1, 2, 3, 4) and $A \cap B$ is (3, 4). Therefore:

$$\frac{|A \cap B|}{|B|} = \frac{2}{4} = \frac{1}{2}$$

Conditional probability is an important concept because, in many situations, the likelihood of one outcome can differ radically depending on how something else comes out. The probability of passing a test given that one has studied all of the material is generally much higher than the probability of passing a test given that one has not studied at all. The probability of a person having heart trouble is much lower if that person exercises regularly. The probability that a college student will graduate is higher when their SAT scores are higher, and so on. For this reason, there are many people who are interested in conditional probabilities.

Note that in some practical situations, changing the order of the conditional probabilities can make the outcome very different. For example, the probability that a person with heart trouble has exercised regularly is quite different than the probability that a person who exercises regularly will have heart trouble. The probability of a person receiving a military-only award, given that he or she is or was a soldier, is generally not very high, but the probability that a person being or having been a soldier, given that he or she received a military-only award, is 1.

However, in some cases, the outcomes do not influence one another this way. If the probability of A is the same regardless of whether B is given; that is, if $P(A|B) = P(A)$, then A and B are considered independent. In this case:

$$P(A|B) = \frac{P(A \cap B)}{P(B)} = P(A)$$

Which means that:

$$P(A \cap B) = P(A)P(B)$$

In fact, if $P(A \cap B) = P(A)P(B)$, it can be determined that $P(A|B) = P(A)$ and $P(A|B) = P(B)$ by working backward.

Therefore, B is also independent of A.

An example of something being independent can be seen in rolling dice. In this case, consider a red die and a green die. It is expected that when the dice are rolled, the outcome of the green die should not depend in any way on the outcome of the red die. Or, to take another example, if the same die is rolled repeatedly, then the next number rolled should not depend on which numbers have been rolled previously. Similarly, if a coin is flipped, then the next flip's outcome does not depend on the outcomes of previous flips.

This can sometimes be counter-intuitive, since when rolling a die or flipping a coin, there can be a streak of surprising results. If, however, it is known that the die or coin is fair, then these results are just the result of the fact that over long periods of time, it is very likely that some unlikely streaks of outcomes will occur. Therefore, avoid making the mistake of thinking that when considering a series of independent outcomes, a particular outcome is "due to happen" simply because a surprising series of outcomes has already been seen.

There is a second type of common mistake that people tend to make when reasoning about statistical outcomes: the idea that when something of low probability happens, this is surprising. It would be surprising that something with low probability happened after just one attempt. However, with so much happening all at once, it is easy to see at least something happen in a way that seems to have a very low probability. In fact, a lottery is a good example. The odds of winning a lottery are very small, but the odds that somebody wins the lottery each week are actually fairly high. Therefore, no one should be surprised when some low probability things happen.

The *addition rule* for probabilities states that the probability of A or B happening is:

$$P(A \cup B) = P(A) + P(B) - P(A \cap B)$$

Note that the subtraction of $P(A \cap B)$ must be performed, or else it would result in double counting any outcomes that lie in both A and in B. For example, suppose that a 20-sided die is being rolled. Fred bets that the outcome will be greater than 10, while Helen bets that it will be greater than 4 but less than 15. What is the probability that at least one of them is correct?

We apply the rule:

$$P(A \cup B) = P(A) + P(B) - P(A \cap B)$$

In this rule, A is that outcome x is in the range $x > 10$, and B is that outcome x is in the range $4 < x < 15$.

$$P(A) = 10 \times \frac{1}{20} = \frac{1}{2}$$

$$P(B) = 10 \times \frac{1}{20} = \frac{1}{2}$$

$P(A \cap B)$ can be computed by noting that $A \cap B$ means the outcome x is in the range $10 < x < 15$, so:

$$P(A \cap B) = 4 \times \frac{1}{20} = \frac{1}{5}$$

Therefore:

$$P(A \cup B) = P(A) + P(B) - P(A \cap B) = \frac{1}{2} + \frac{1}{2} - \frac{1}{5} = \frac{4}{5}$$

Note that in this particular example, we could also have directly reasoned about the set of possible outcomes $A \cup B$, by noting that this would mean that x must be in the range $5 \leq x$. However, this is not always the case, depending on the given information.

The *multiplication rule* for probabilities states the probability of A and B both happening is:

$$P(A \cap B) = P(A)P(B|A)$$

As an example, suppose that when Jamie wears black pants, there is a $\frac{1}{2}$ probability that she wears a black shirt as well, and that she wears black pants $\frac{3}{4}$ of the time. What is the probability that she is wearing both a black shirt and black pants?

To figure this, use the above formula, where *A* will be "Jamie is wearing black pants," while *B* will be "Jamie is wearing a black shirt." It is known that $P(A)$ is $\frac{3}{4}$. It is also known that $P(B|A) = \frac{1}{2}$. Multiplying the two, the probability that she is wearing both black pants and a black shirt is:

$$P(A)P(B|A) = \frac{3}{4} \cdot \frac{1}{2} = \frac{3}{8}$$

Lagging Indicators

In occupational health and safety (OHS), **lagging indicators** are measurable factors that refer to the frequency of company incidents, such as illness, injuries, or fatalities, based on past statistics. These indicators are safety metrics that are used to show progress toward safety rule compliance. Some examples of lagging indicators include injury severity, recordable OSHA injuries, workers' compensation costs, and lost workdays. The overall effectiveness of safety at a workplace is determined by these indicators. Although these indicators may measure effectiveness by indicating how many people are injured at a workplace, they don't reflect how well a company is trying to prevent accidents or incidents.

Incidence Rates and Lost Time

Construction establishments with eleven or more employees (on any day of the calendar year) are required to record employee injuries and illnesses on the appropriate OSHA recordkeeping forms. OSHA's recordkeeping standard can be found at 29 CFR 1904.

OSHA does not require every injury and illness to be recorded; only incidents that meet one or more of a predetermined list of severity or event criteria are required to be entered on the OSHA recordkeeping logs.

Because of the importance of accurate incident rates and the data from which the rates are derived, an ASP must commit to memory the particular events that will automatically trigger OSHA recordkeeping and the various types of cases.

All deaths must be recorded, with the exception of "natural causes" and preexisting health conditions and events such as heart attacks. A medical examiner or other licensed medical professional will need to make the determination that the death was not work-related to avoid log entry.

The employer must record a case if an employee misses one or more days away from work or is assigned modified duty following the day of the incident. Modified duty includes work restrictions and transfers. Examples of modified duty are partial workdays, lifting restrictions, and not performing routine job functions. Transfer duty is whenever an employee is temporarily assigned to another position; for example, a construction laborer could be transferred to the office to help with clerical tasks while a back sprain heals.

The OSHA Form 301 is the Injury and Illness Incident Report. This form is used to record basic details of an incident, such as employee information, day and time of the incident, and the particulars of the incident. The 301 must be completed within seven calendar days of the incident.

The OSHA Form 300 is the Log of Work-Related Injuries and Illnesses. This is essentially a basic spreadsheet for logging OSHA-recordable injury and incident events. Data from the OSHA Form 301 that identifies the person affected and a brief description of what happened will provide the background for the OSHA 300 log entry.

Each incident case is then classified on the OSHA 300 log by the most serious event outcome for that particular case. There are four columns for recording the four classifications/outcomes in order of seriousness, the first of which is death. **Days away from work cases** are the most serious nonfatal outcome. The next most serious incidents are **job transfer or restriction cases** that involve modified duty.

For incidents resulting in missed workdays or modified-duty days, only days following the day of the incident itself factor into the classification. Assume that an employee is cut in the morning, receives stitches at noon, and then returns to work prior to the end of the workday, or perhaps even the next shift in the morning. This case will not be classified as involving days away from work because only days following the day of the incident are counted for this purpose; the same treatment applies to days with modified duty.

There will be times when an incident case involves both days away and modified-duty days. With these cases, it is important that the case be recorded as a single log entry and that it be classified as only one type. Since OSHA's recordkeeping scheme identifies days away from work as a more serious outcome than days with job transfer or restriction, such cases should be recorded only in the column for **days away from work cases**. If such a case is classified as both types, then at the end of the year when the column entries are added up, the total number of cases will be one more than what actually occurred.

Other recordable cases are the fourth and least serious event outcomes. These incidents do not result in any days away from work or modified-duty days but are still recordable because one or more severity thresholds have been met, for example, a loss of consciousness (even for just a minute), medical treatment, prescriptions, and significant injuries and illnesses.

At the end of the year, the total numbers of the four incident types that occurred will be entered on the OSHA Form 300A: the Summary of Work-Related Injuries and Illnesses. Establishment information, the average number of employees, and the total hours worked by all employees in the reporting year are also entered on the 300A. The 300A must be completed by February 1 of the next year and posted in the workplace from February 1 until April 30.

The establishment's Northern American Industrial Classification System (NAICS) code is also entered on the 300A. The NAICS code is a number of up to six digits that identifies the type of business activity that an establishment undertakes. All construction employers have an NAICS code beginning with 23.

The data found on the OSHA 300A can be used to calculate the establishment's incident rates.

The "Days Away" rate is based on the number of cases involving days away from work following the day of the incident and is also referred to as DAFWII (Days Away From Work Injured and Ill), LWDI (Lost Workday Incidents), and LTIR (Lost Time Incident Rate).

The DART (Days Away, Restricted, and Transferred) rate is based on the number of cases involving days away from work following the day of the incident and also the cases involving days of restricted and transferred duty.

The TRC (Total Recordable Cases) rate is based on the number of all recordable cases in the calendar year and is also referred to as TRIR (Total Recordable Incident Rate).

The total recordable incidence rate (TRIR) is:

$$TRIR = \frac{\text{number of OSHA incidents per year}}{\text{hours worked by employee per year}} \times 200{,}000$$

A lost-time injury incidence rate (LTIIR) refers to the days an employee missed work due to an injury. The LTIFR is expressed using the following formula:

$$LTIIR = \frac{\text{work days lost per year}}{\text{hours worked by employee per year}} \times 200{,}000$$

The value of 200,000 refers to 100 employees who worked for 50 weeks with 40 hours per week ($100 \times 50 \times 40 = 200{,}000$).

Direct and Indirect Costs of Incidents

Workplace injuries include direct and indirect costs. **Direct costs** refer to the worker's medical expenses, compensation, and legal services. **Indirect costs** include lost productivity, repairs to damaged property, accident investigation, training replacement employees, and employee morale.

Leading Indicators

Leading indicators are preventive measures used to prevent and control an injury; they focus on continuous workplace improvement and future safety performance. Some leading indicators include safety training or interventions, ergonomic opportunities, employee perception surveys or performance evaluations, and safety audits or inspection frequency. Leading indicators can improve company performance by preventing workplace injuries/illnesses, reducing costs connected to incidents, improving productivity, optimizing health and safety performance, and increasing/raising worker participation.

Inspection Frequency

Inspection frequency refers to how often a system or piece of equipment should be subject to an inspection. Inspection frequencies vary. Relevant factors to consider for frequency include, but are not limited to, the type of system or equipment, time, practical considerations, and desired goal. The desired goal must be identified, and measurable stages may be implemented to achieve that goal. Reviewing prior information and data may show when an inspection is needed to deter undesired conditions. By engaging in these objective checkpoints, it allows for an entity to engage in conduct that will help obtain the desired goal (or performance). Inspections for equipment may be performed more frequently, such as inspecting a semi-truck after transporting materials. However, inspections of implemented management procedures may not occur as frequently because enough time has not yet passed to provide relevant information toward achieving a desired goal, such as an inspection of annual performance.

Safety Interventions

Safety interventions include safety and health programs that are aimed to prevent workplace injuries, deaths, and illnesses. The programs will also attempt to prevent financial hardship and suffering that may be imposed on workers and their families. Safety interventions will encourage practices that will find and fix hazards before they result in injury or death. Safety programs start by achieving simple goals that include monitoring performance and evaluating outcomes so that the workplace can attain higher levels of safety. The advantages of safety interventions include preventing workplace injury, improving compliance with the law, reducing workers' compensation costs, engaging workers, clarifying goals, and increasing productivity.

Employee Performance Evaluations

The **International Organization for Standardization (ISO)** is known world-wide for innovation in the standardization industry. It created a system known as ISO 45001. It is similar to the structure of OHSAS 18001, but it concentrates on risk management while OHSAS 18001 focuses more on hazard management. Under this system, ISO has developed an evaluation of employee performance. It utilizes a process approach and the PDCA Continual Improvement Cycle. A process approach allows for gradual stages to occur so that a desired management goal may eventually be met. The PDCA is a continuous cycle that involves planning, doing, checking, and acting to ensure that there is improvement toward a desired management goal. Combined, these two methods create a system to evaluate employee performance and ensure that management goals are being met.

The evaluation is a critical self-examination that shows how to collect data, interpret the data, and how management should address the results. An internal audit then verifies these results. This evaluation includes previously determined legal (and other) requirements and ensures that there is compliance with regulations.

Training Frequency

OSHA mandates that employees are trained after hiring and throughout the year. **Training frequency** refers to the frequency or number of times an employee is given workforce training. Employees are given training after hiring and anytime the emergency response plan changes. Training will depend on the work environment and will cover the appropriate response to a specific emergency. Training for emergencies will cover the actions needed to protect employees; emergencies may include unexpected chemical contamination, natural disasters, fires, and injuries resulting from the incorrect use of PPE. OSHA regulations specifically outline the training frequency for specific emergency responses.

Near-Miss, Near-Hit, and Close-Call Reporting

Near-miss or near-hit, also called **close-call reporting**, refers to an unplanned event that doesn't result in damage to equipment/environment or illness/injury to a person. Depending on the situation, near-miss events may be considered either a lagging or leading indicator. For instance, if near-miss events track worker injury, they may be considered lagging. If near-miss events track potential injuries/illnesses that haven't occurred, they may be considered leading.

Hierarchy of Hazard Controls

Hierarchy of Controls

Hierarchy of controls is a long-established protocol for assessing hazards and prioritizing workplace mitigation measures for such. It is necessary to reconcile the mitigation of worksite hazards with the resources that are available and/or allotted to this end. Practical considerations dictating every hazard cannot be completely removed from a worksite, and therefore consideration must be given to prioritizing the hazards with the highest potential harm to workers. The hierarchy of controls is used in concert with a **cost-benefit analysis (CBA)** of worksite hazards to determine resource allocation where they are perceived to be needed the most. The hierarchy of controls is one of the primary tools utilized in the NIOSH initiative for **prevention through design (PtD)**. Based on the concept of designing hazards out of the workplace, PtD applies preemptive design changes to work methods, tools, equipment, operations, processes, worker facilities, newer technologies, and workplace organization. The hierarchy of controls includes the following:

- **Engineering**: This covers the middle ground between elimination/substitution and administrative/PPE. It is a method whose purpose is to remove a worksite hazard at its source prior to worker exposure and by design. It is also adaptable in an already-existing work site. An engineering control generally consists of modifications or barriers designed into a worksite that preclude workers' exposure to hazard. Removing a source worksite hazard is, in respect to the **corrective and preventive action (CAPA)** process, a preventive action. Identifying, mitigating, and eliminating a hazard before it becomes an incident is a proactive, preventive measure.

- **Administrative**: These controls tend to be less expensive than other mitigation measures but are significantly more reliant on worker participation, as they are implemented in areas with decreased hazard controls in place. HSE signage, postings, and training are easily and inexpensively implemented, although more challenging to incorporate. Administrative controls must be part of the CAPA process, as they are required to guide personnel assigned to corrective and preventive tasks in order to effectively implement the CAPA plan on file.

- **Personal protective equipment**: As with administrative measures, PPE is used in areas where hazards are not very effectively controlled. Heavily dependent on monitoring, mentoring, and worker participation, this program is contingent on effective and consistent worksite monitoring by personnel such as the ASP. Standard PPE includes boots, a hard hat, gloves, and safety glasses, but can include other equipment specific to the task at hand. As part of the CAPA process, PPE can be considered both preventive (if used proactively before an incident occurs) and corrective (if used in correction after a hazard has been identified or post-incident).

- **Substitution**: Substitution, while one of the preferred hazard mitigation methods, typifies the more desirable but higher-cost hazard solution. It becomes increasingly difficult and less cost-effective to implement substitution controls once the worksite has already been established. When the contractor chooses the means and controls for a work project, worksite equipment, materials, and tools have usually been ordered and delivered. Changing any of these resources during an already established project or altering the worksite can be prohibitively expensive.

However, if an identified hazard exists, or an incident has occurred that has resulted in financial or human loss, substitution may be the only solution in a corrective action.

- **Elimination**: This is by far the most preferable in an ideal environment and typically the least flexible and most expensive of controls. The optimum time for the application of elimination controls is during the project planning stage while it is still feasible to make major changes. Once work is under way, it is more expeditious to default to administrative or PPE controls than to implement changes using the elimination control. Elimination should be considered a preventive action in the CAPA process.

Using Hierarchy of Controls to Protect Workers

Applying the **hierarchy of controls** to protect workers requires a specific logic and behavioral flow. This can be conceptualized via:

- **Higher-level controls**, which focus primarily on design, equipment, and materials *at the source.*
- **Mid-level controls**, which seek to reduce existing hazard risk through system redesign and process modifications.
- **Low-level controls**, which protect workers via modifying behaviors and reducing hazard impact *at the point of the worker.*

The most effective means of eliminating or reducing risk requires the understanding that all higher-level controls should be analyzed and exhausted prior to considering lower-level controls because the highest-level controls more greatly affect hazard presence and risk, while the lowest levels affect the consequence of existing hazards. The following includes a prioritized list of hierarchical controls to protect workers.

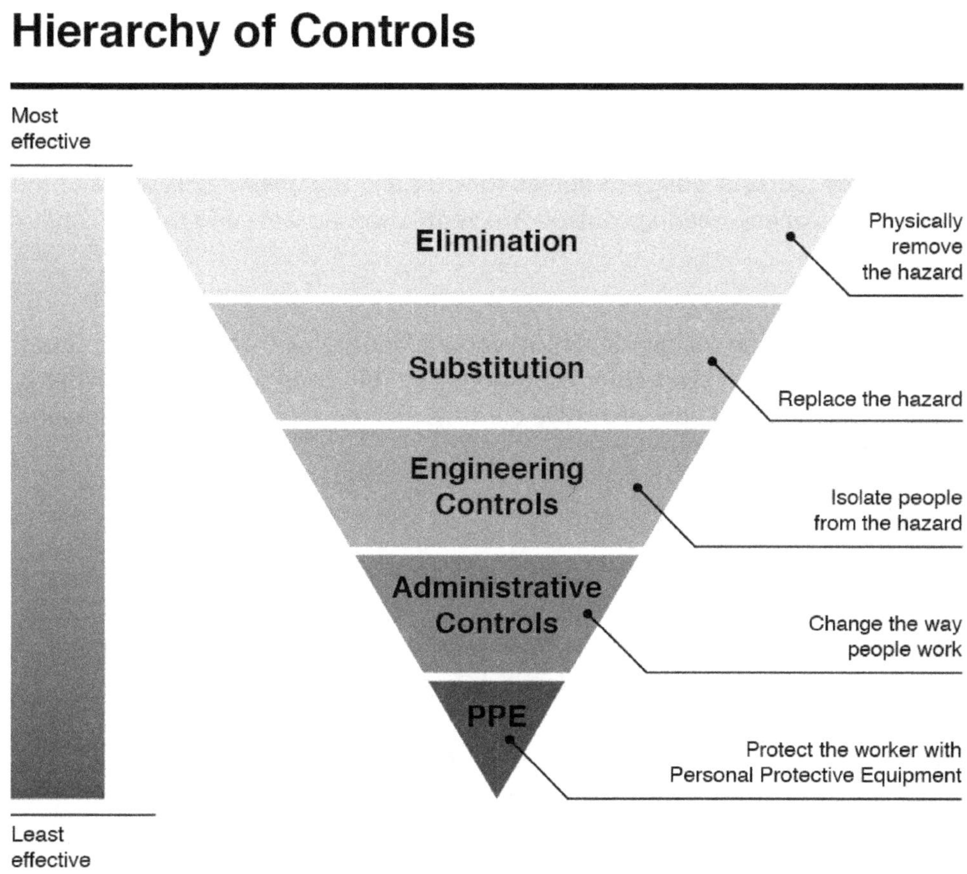

I. Safety Through Design (Aspect of Elimination)

Applying safety through design is the best means of protecting workers as hazard risk is most effectively addressed at the lowest cost. Eliminating hazards here greatly enhances an organization's control of processes, provides worker protection, and reduces risk assessment efforts subsequent to project start-up.

II. Redesign (Aspect of Elimination)

This stage of action requires design and equipment modification to address system hazards. Redesign proposals often present issues with practicality, feasibility, and additional costs. Elements of redesign can include measures such as process automation and extensive fall-prevention systems.

III. Substitution

Substitution involves removing a hazardous process or material and replacing it with a less hazardous alternative. Substitution can be effective for risk reduction, but its reactive nature (post-incident), higher residual risk, and compounding costs make it less effective than design and redesign measures. Examples of substitution include replacing a hazardous chemical with a more inert alternative and introducing a powered conveyor to mitigate risks associated with manual material handling.

IV. Engineering Controls

Engineering controls are components introduced into a system to isolate an existing hazard. This reduces hazard risk by separating workers and processes from the hazardous element. Relative concerns include workers' ability to defeat controls and the ongoing presence of the hazard source. Examples of engineering controls are ventilation systems and machine guarding.

V. Administrative Controls

These controls affect the manner in which workers behave or react to a given situation. They involve avoiding uncontrolled hazard conditions and rely upon worker capabilities and human performance. Examples of administrative controls include selecting specific personnel to perform a task and training workers.

VI. Personal Protective Equipment (PPE)

Although PPE is vital for protecting workers against injury and illness, it is the least effective method of control in the hierarchy. PPE places the element of protection, or barrier, at the point of the worker, can negatively affect performance, and can easily be defeated. PPE comes in an array of forms including face shields, gloves, boots, hearing protection, etc.

Risk Transfer

Insurance

Workplaces, especially construction sites, present a certain amount of risk from injury, exposure to chemicals, budget shortfalls, missed deadlines, et cetera. The first step in mitigating risk to avoid claims is to identify and assess potential risks to better plan and manage them.

Mitigating risk means reducing negative impact by avoiding, limiting, transferring, or accepting the risk. The process of mitigating risk includes identifying hazards in the workplace that could produce negative effects; assessing the likelihood and impact of the risk in the event of exposure; identifying ways to limit exposure or impact; removing hazards that present risks or implementing strategies to control exposure; and devising fallback plans or contingencies if exposure is unavoidable.

Depending on the type of industry or the nature of the project, hazards on the job site may present a risk to public safety. Risk management should include strategies to communicate with the public in case of exposure and respond efficiently to limit negative impact and protect the safety of the public.

Builder's risk insurance covers damage to property that is under construction. This policy protects the builder's interest while they are working on a property to ensure they are not liable for losses caused by natural disasters, fire, or high winds. It also ensures the builder is not liable for damage that occurs during the course of the construction work. It does not cover flood or earthquakes. Builder's risk coverage ends when the construction is completed. Such a policy can be purchased by the contractor (builder), the property owner, or both.

Builder's risk coverage applies to the property, as well as the equipment and the materials used to complete the construction work, but it does not cover injuries to workers or protect contractors or property owners from claims by workers or third parties if an injury occurs on the property during the construction work. General liability insurance covers these types of losses from accidents on the job or claims against the contractor or property owner if someone is injured by the completed work. General liability does not cover claims due to faulty workmanship or negligence.

Value engineering is an estimation method used to find the most cost-effective project construction processes without compromising worker safety; this makes the accuracy of the initial hazard identification and the ongoing hazard identification methods such as the JSA even more important. A JSA is a mini risk assessment that enables the workers charged with a task to participate in the entire safety vetting process while providing a daily sounding board and worker interaction. This is always a great opportunity for the ASP to mentor in the application of the key risk management elements: identify, assess, analyze, manage, monitor, and control. The ASP has the capability to inspire a meaningful exchange between workers during a JSA. To achieve this aim, the ASP must effectively encourage expressions and ideas from the most reticent workers while arbitrating the discussion to rein in the more outspoken ones. To balance both of these elements, the ASP is obliged to listen more and to choose speaking points wisely and economically.

Risk Tracking Flow Diagram

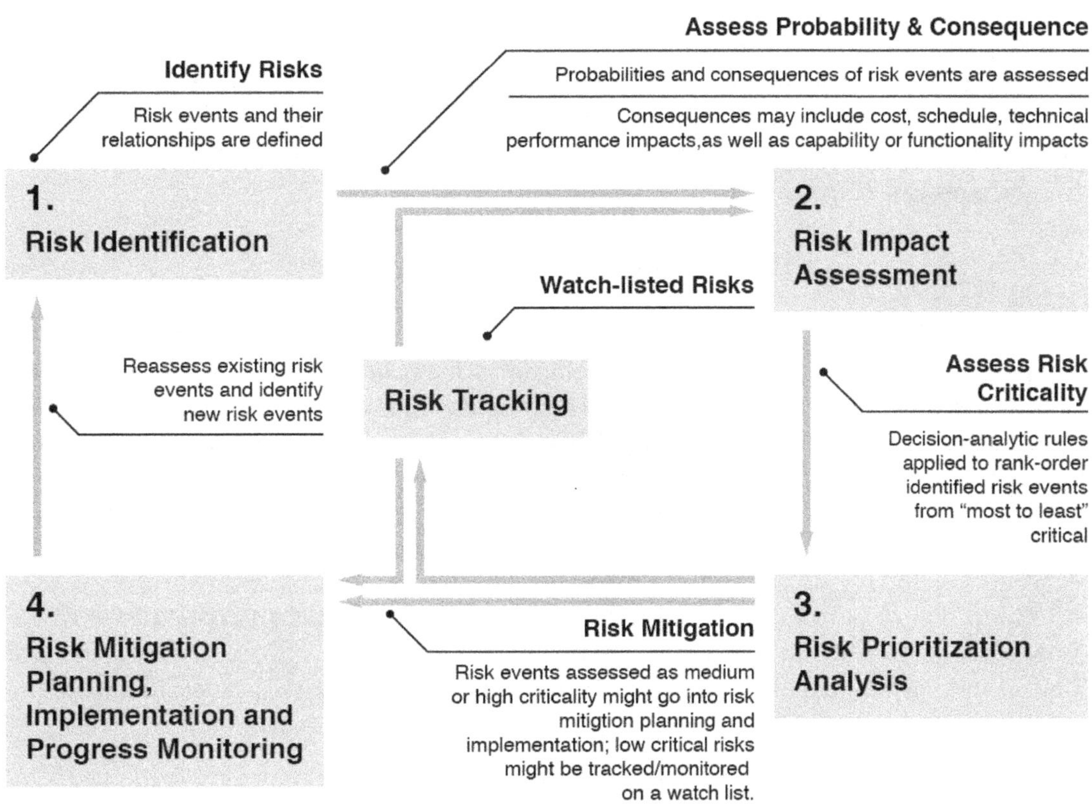

Outsourcing

Outsourcing, also referred to as risk sharing or risk transfer, is the practice of allocating a service or duty to an external third party. Outsourcing can be a great tool for transferring a specialty or unknown subject matter to a qualified entity. For example, a construction company may choose to outsource its IT development and maintenance to an IT firm. This would allow the construction company to have sufficient IT structure without designating the duty to individuals that may not be aware of special processes and practical procedures that go into such a duty. It is common to designate an external entity that is near the company's location so that issues, such as onsite service, may be addressed quickly and efficiently.

Like other decisions and risks, there are costs and benefits associated with outsourcing. Such benefits may include access to efficient service, qualified personnel, lower costs, specialized equipment, innovative practices, and other monetary benefits. However, costs may include prolonged completion of projects, lack of industry knowledge by the company used for outsourcing, lack of control, and communication barriers.

Outsourcing services is a term that refers to outsourcing a specific duty. Outsourcing services may be beneficial, but it is important to determine a qualified entity that may be retained for the outsourced duty. Decisions regarding an outsourcing service should be based on objective research and attenuating circumstances. Using a cost benefit analysis may be helpful in determining whether to outsource and where to outsource. Determining the appropriate proper external third party may be difficult, but the above findings will help corroborate whether to outsource a duty.

In the event that a company is subject to liability from an outsourced obligation, the company may have no liability or limited liability on the matter. It depends on relevant agreements and conditions regarding the disputed matter. Either the external entity may be wholly liable or subject to an indemnification agreement with the company. Likely the best way to handle outsourcing is to have a service agreement that is clear on the duty that is being outsourced, and the company's expectation regarding the agreement and events that may arise.

Management of Change

Health and safety programs in construction must consider ongoing efforts to manage change. This adjustment process prevents new hazards from being introduced into the work environment. New hazards can stem from changes in equipment and technology, new design specifications, different materials used, changes to personnel affecting skill and experience, altered procedures, and modified standards and codes. An objective of change management is to ensure proper program maintenance with timely revisions and updates. Implementation and revisions must account for the following:

- Measures do not create an entirely new set of hazards.
- Measures do not increase the frequency, severity, and risk level of an existing hazard.
- Previously eliminated hazards are not reintroduced.
- Conceivable risk should be analyzed, and unforeseen/unexpected risk level should be minimized.

Management of change refers to the practices and means that a company utilizes to implement change. Changes may be made to internal or external processes. Generally, this requires providing employee support, establishing necessary stages of change, end evaluating conduct and performance prior to and after the change. Some organizations may view change as overwhelming, but implementing a system for management of change allows for a smoother transition. It is imperative that change is implemented in stages of development. The different stages of development must engage multiple levels of the organization, and sometimes external entities are involved in the transition. These stages will help provide structure to the transition while reducing the likelihood of potential issues.

As discussed, change can be overwhelming, and this experience may cause change to fail. Reasons for failure include, but are not limited to, lack of supervisor involvement, lack of employee openness to change, and lack of communication of expectations. Communication is an effective tool that will allow for change to be smoothly implemented across the organization. It is helpful for all segments within the organization be aware of the desire for change and what the anticipated result of the change will be. Different segments within the organization are affected differently by change, and communication will allow all segments to objectively see how change is affecting the organization overall.

Prior to implementing change, the changes to be made must be determined. Additionally, this determination should identify the team to manage the change and develop the various stages of change. After the tentative plan has been developed, it must be implemented. The plan may not cover all issues relating to the change, but the plan (and team) should be open to flexibility within the management plan and their practices and procedures.

It is not uncommon for an organization to face resistance to change. Resistance may be from internal traditional views, external pressures, or other industry related matters. Resistance can also be related to the existing systems, processes, or equipment of the organization. This resistance is a reinforcement of why the developed management of change should be flexible in order to address these issues and ensure a smooth transition.

Generally, management of change will have three stages: (1) prior to the change; (2) during the change; and (3) after the change. Prior to the change, organizations should determine the desired change and raise awareness with the necessary internal or external entities. This awareness should create an understanding and affirm why this change is being implemented and what the anticipated result is. During the change, there must be involvement from supervisors and employees to ensure that change is being implemented safely and efficiently. This involvement will allow organizations to measure implemented changes as well as ensure that the desired change is being obtained. The objective measurements will show that an organization is proceeding as desired. After change has occurred, the desired goal should have been met, and an organization is able to evaluate its performance under the change. These performance evaluations will be recorded, and the organization will be able to refer to the data at a later time.

Hazard and Risk Analysis Methods

Preliminary Hazard Analysis

Preliminary Hazard Analysis (PHA) is a tool used to determine potential hazards that may cause harm and is categorized based on severity. It is used to identify applicable regulations and necessary follow-up to an incident. Generally, this is created in the primary stages of a project. The PHA should establish a team that is competent and able to implement PHA procedures. The size of the team is contingent on the complexity of the subject matter. The team must be familiar with the subject matter and potential hazards that may occur. Additionally, it should be known how severe a hazard may be, what types of incidents may stem from the hazard, and how often (or likely) an incident is to occur. A strong PHA is one that is built on experiences and previous history of the subject matter and procedures that are the likely best for handling potential incidents.

Subsystem Hazard Analysis

Subsystem Hazard Analysis (SSHA) is an analysis of the system to determine its ordinary conduct, function duration, and potential functional issues that may occur. The purpose is to evaluate the system's construction and integrity and note its limitations. The comprehension of the system's subsystem design allows for a greater understanding of how to optimize the system's conduct and function under the surrounding circumstances and facts. This analysis provides a practical perception of what is likely to occur from the system when it is used for its ordinary purpose. However, this analysis does not include prior unidentified hazard system conduct.

Hazard and Operability Analysis

Hazard analysis is an initial step that focuses on work tasks to assess risk or identify the different types of hazards that can occur. The OSHA 1926 standard requires a hazard analysis and lists methods that can be employed. The methods for evaluating hazards include the following.

1. A what-if, checklist, and fault tree analysis (FTA)
2. A failure mode and effects analysis
3. A what-if and checklist assessment
4. A hazard and operable study or methodology

In reality, a certain amount of risk must be accepted and is generally determined by management. The expected costs of not implementing hazard controls, or accident costs, must be compared to the marginal costs of implementing hazard controls. Two factors must be considered to quantify accident costs. The first factor relates to the risk or consequences of an accident, and the second factor refers to the probability of an accident taking place. For instance, suppose an employee works in an area that contains toxic gases. Without protection, exposure to the gases will cause severe lung damage. Because the severity of the injury or probability of an accident is highly probable, there are high risks or consequences; the risk of not implementing a hazard control is unacceptable.

The severity of an accident or injury may be quantified in dollars. Therefore, the costs of not implementing hazard controls will be greater than spending money to implement a hazard control, such as personal protective equipment (PPE), for an employee. A high accident probability means the risk factor is unacceptable; the company is more likely to spend money to implement a hazard control. A low accident probability means the risk factor is acceptable; the company is less likely to spend money to implement a hazard control.

It's beneficial to spend more money on implementing hazard controls to reduce the probability of the accident. For accidents that have less severe consequences, there can be a larger probability of occurrence that is acceptable, and less money will be spent to reduce the frequency of these types of accidents.

Failure Mode and Effects Analysis (FMEA)

Failure Mode and Effects Analysis (FMEA) is an approach used to identify all the possible ways that a design, assembly, or manufacturing process, service, or product can fail. FMEA also tries to identify the associated consequences of such failures. These failures are given prioritization based on their ease of detection, the seriousness of their consequences, and the frequency with which they occur. Beginning with the highest priority failures, the goal of FMEA is to eliminate or reduce them.

It's fitting to use FMEA when a product, service, or process is being designed or redesigned, or when an existing product, service, or process is being applied in a new manner. In addition, FMEA can be helpful when analyzing the failures of an existing product, service, or process, or before control plans are developed for a modified or new process. FMEA can also be beneficial when improvement goals are outlined for an existing product, service, or process.

Fault Tree Analysis

A certain undesired event, such as a postulated accident condition, can be studied analytically using a deductive tool called a **fault tree analysis (FTA)**. All known events, faults, or occurrences that contribute to that accident must be known in order to use an FTA. In the context of engineering, FTA was used to enhance the safety of missile systems. Within a system, most accidents will result from failures. The system can contain environmental factors, equipment, and people. Because environmental factors and the system are interconnected, a failure in one component will affect other components. Analytical trees are graphical representations of an event that use deductive reasoning; they start with an output event and move down through branches, like a tree, to specific input events that must take place. Fault trees are negative analytical trees that serve as troubleshooting tools. Fault trees can prevent failures before they occur. When an accident occurs, the main cause of that negative event can be pinpointed. Negative events are incidents that can result in property damage or personal injury.

Fishbone

Cause-and-effect diagrams are used for analyzing all potential causes or influences for a given effect or problem in various organizational situations. Also referred to as fishbone diagrams, they can provide a clear, visual representation of the factors affecting a system or process. Both positive and negative effects can be analyzed using cause-and-effect diagrams.

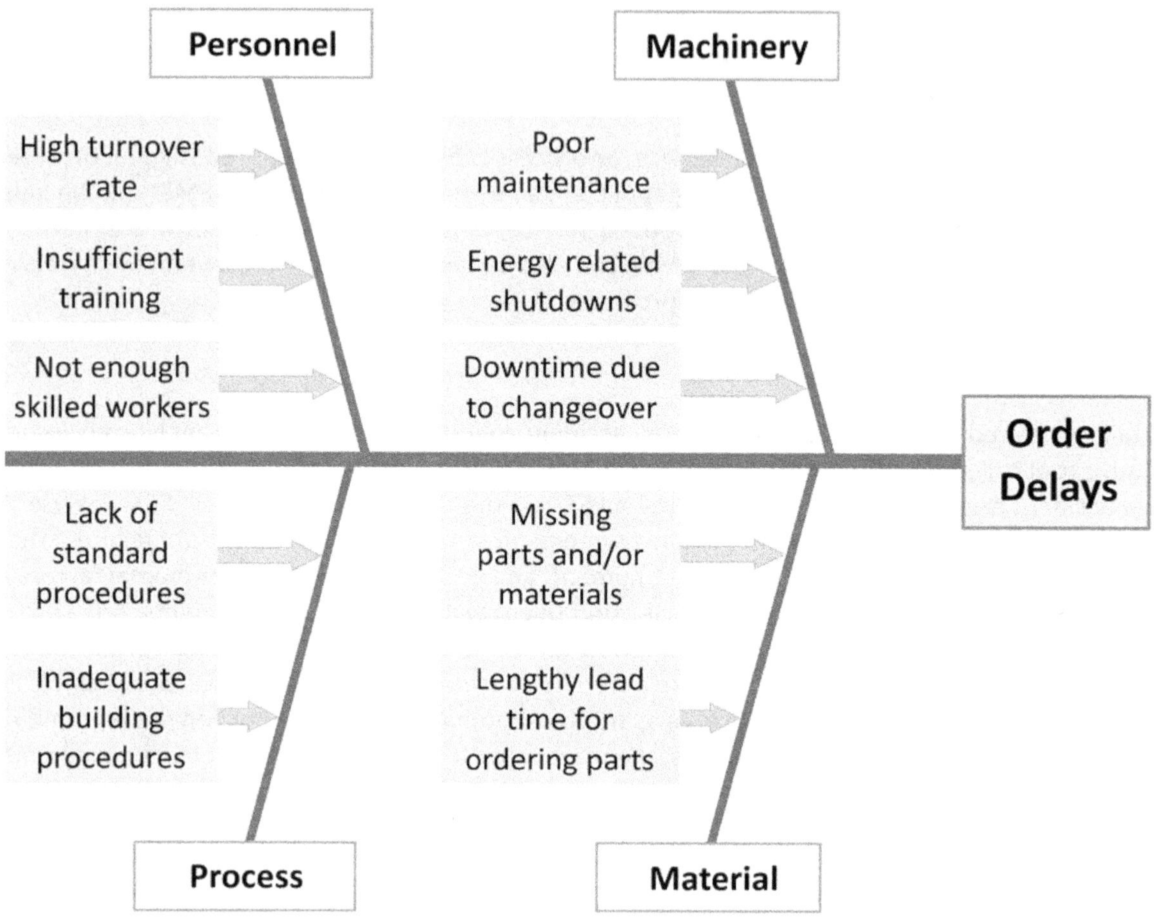

The above illustration represents a sample cause-and-effect diagram analyzing factors contributing to delays in custom orders. Four likely factors (material, personnel, process, and machinery) are further examined to create branches and brainstorm all potential causes. The fishbone diagram is a simple yet powerful tool that can be provided to management and other relevant teams.

What-if/Checklist Analysis

What-if/checklist analysis combines two methods. The **what-if method** is carried out by a team with diverse backgrounds that include safety and engineering; using critical-thinking methods, the team looks at a specific process and creates what-if questions. For example, a process may involve the use of hazardous chemicals or mechanical equipment and ask, "What if a water pump fails?" The team lists likely hazards and recommends actions to prevent unwanted outcomes. The **checklist analysis** consists of the following:

1. The complete industry experience for a process

2. The past incident and accident experience

3. A detailed description of the technology process

4. Listing the known hazards for a process and the associated materials/chemicals

5. Listing the outcomes/reviews of an existing hazard

6. Reviewing piping and instruction diagrams (P&IDs) with instrument and control devices

Change Analysis

Change analysis is a method of solving problems by observing a process that is not performing efficiently. The process may not be working properly, or at all. The analysis compares the non-functioning process to a functioning version of it. The side-by-side comparison allows for an assessment of what components are common between the two and what is different. The differences help identify potential causes for the performance issue in the process. By reviewing and evaluating the differences, it shows which difference is the actual cause of the issue at hand. When addressing the issue, the non-functioning process should be operating under its intended function; however, it may take multiple attempts to determine the cause.

Energy Trace and Barrier (ETBS) Analysis

Energy Trace and Barrier (ETBS) Analysis is a process used to determine hazards by evaluating the energy associated with a system and the barriers that are used for the energy's regulation. This analysis shows an in-depth perception and comprehension of a system's energy transference. It is an understanding of the energy flowing into a system, where the energy flows internally, and how the energy affects the system's design and performance. The tracing of the energy shows the conduct and performance of a system under particular circumstances or events that are being evaluated.

Systematic Cause Analysis Technique (SCAT)

System Cause Analysis Technique (SCAT) is used to develop a chart that shows components of a loss event. The purpose of the chart is to evaluate loss and determine what matters may be addressed to reduce loss going forward. The categories are:

(1) loss that occurred;

(2) event that occurred;

(3) direct cause;

(4) basic cause; and

(5) lack of controls.

The loss that occurred is the negative result affecting a person or property. The event is the behavior that resulted in the loss occurring. The direct cause is the surrounding facts and circumstances that initiated the event. The basic cause is the preceding cause that initiated the direct cause – this can include such things as incompetent (below standard) personnel or property that was operating at the time. The lack of controls refers to not having preventative measures in place to prevent the loss event.

Process Safety Management

Injuries and illnesses are a burden for both employers and employees. The loss of productivity due to workplace injuries and illnesses can be significant, and the loss of income for employees can affect an individual and economy as a whole. Therefore, organizations must establish programs that minimize or prevent these incidents. If a workplace injury or illness does occur, workers' compensation may provide fixed payments to the employee. Workers' compensation also covers dependents of those who are killed as a result of workplace accidents. Limits do exist for these compensation benefits, such as caps on what can be collected from employers.

OSHA, a federal agency that is designed to ensure safe working conditions for employees, established process safety management standards that deal with hazardous chemicals in the workplace. If an employee could potentially come into contact with the hazardous substance during the normal course of their jobs, these substances must be properly evaluated, classified, and labeled. This information is recorded in material safety data sheets (MSDS), which must be easily accessible to individuals who work with any hazardous materials. The MSDS should state what should be done if someone has inappropriate contact with the chemicals, such as an employee who splashes a dangerous chemical in their eye.

OSHA has developed standards for employee personal protective equipment (PPE) in hazardous working environments. These items may include safety glasses, hard hats, and safety shoes. Employees are provided these items at no cost and must be paid their rate of pay for the time required to put on and take off protective equipment.

OSHA has established guidelines to assist employers in the event of a pandemic disease outbreak by utilizing proper safety equipment and procedures. The guidelines are also meant to assist the company to continue operations with a reduced workforce.

Ergonomics, or the study and design of the work environment to address physical demands placed on employees, is yet another area addressed by OSHA. In the workplace, ergonomics deals with elements such as lighting, placement of controls, equipment layout, and fatigue. OSHA examines work-related injuries that result from repetitive stress and repetitive motion, such as carpal tunnel syndrome. These are also known as cumulative trauma disorders. These workplace injuries may be reduced by redesigning workstations and improving workplace environments.

Fleet Safety Principles

Driver Behavior

The US Department of Transportation (USDOT) Federal Highway Administration states that drivers average approximately 13,500 miles per year, and the National Highway Traffic Safety Administration (NHTSA) indicates a 7% chance of a vehicle collision. Fleet drivers average close to 25,000 miles per year or more and therefore have a greater chance of collision; the annual accident rate for a fleet vehicle is approximately 20%. Fleet vehicle crashes comprise most worker injury claims with an average greater than $21,000 per incident. These accidents result in lost revenue due to missed sales calls and a loss in productivity; there may also be third-party liability claims. Companies will take on the burden of these costs and consequently should take a proactive stance on fleet safety.

The life of a company vehicle can be extended by implementing programs that improve and teach drivers defensive driving skills in addition to awareness of different vehicle types on the road. Fleet safety programs that offer training to fleet drivers are expected to lower maintenance and accident costs.

Distracted Driving

Distracted driving is an activity that removes the driver's awareness from the road. These activities include talking with passengers, eating, and talking or texting on cell phones. Adjusting vehicle controls, such as an air conditioning (AC) knob, is also a primary safety threat. Distractions may also be mental, such as daydreaming. According to the Insurance Institute for Highway Safety (IIHS), many states have addressed cell phone distraction. Since 2020, the District of Columbia and 24 states have banned talking on cell phones while driving. Texting is banned in the District of Columbia and 48 states.

Defensive Driving

The Safe Practices for Motor Vehicle Operations (SPMVO, ANSI/ASSE Z15.1) defines **defensive driving** as a practice that involves driving with the intent to save lives, money, and time with the consideration of the surrounding conditions and actions of others. Drivers must apply strategies that reduce the risk of a vehicle accident. Fleet driver training must provide instruction in the safe, proper, and efficient handling of vehicles. Hands-on training and written examinations are part of the training. The National Safety Council (NSC) states that it is more effective to give extensive and systematic training to a driver after they are hired. Remedial and ongoing training should be given to drivers to ensure they are enacting safety practices/ideas. Monitoring of driver behavior includes a ride-along with evaluations, an onboard computer monitoring system, and hotline numbers ("How am I driving?"). Defensive driver training should emphasize that the driver should continually recognize hazards and practice space/speed management and vehicle handling. In fact, the lack of appropriate hazard training is one of the most frequent citations given by OSHA and is second only to fall protection in construction. Driver training programs are focused on the following areas.

1. Direct/indirect costs and safety performance measures

2. Appropriate hazard materials training

3. State/municipal driving regulations with an emphasis on Federal Motor Carrier safety for interstate commerce

4. Accidents due to mechanical conditions or defects, lighting/weather conditions, acts of pedestrians, other drivers, and the driver's mental, physical, and emotional state

5. Specific defensive driving for vehicles such as vans, buses, and medium to large trucks

6. Defensive/prevention measure of two-vehicle collisions (vehicles can impact at six different positions)

7. The cumulative costs of an accident when backing up or backing around corners/driveways

8. The importance of maintaining a safe distance (stopping distance), driving maneuvers, driving in traffic, procedures in case of an accident, and operating procedures for a safety program

Fatigue

Fatigue is a condition felt by a person that requires sleep. The condition is the body's response to not receiving sufficient quality sleep. It hinders a person's ability to conduct matters safely, especially driving. While driving with fatigue, a person may nod off, react slowly to surrounding conditions, and experience other symptoms that create a dangerous situation for surrounding persons and property. Additional symptoms may occur, based on severity. The best preventive to fatigue is quality sleep. Companies must create procedures and policies that regulate how many hours a person may work in a determined time. The procedures and policies are designed to ensure that employees have the opportunity to get sufficient sleep. Additionally, there must be sufficient staffing within an entity to ensure that employees are not constrained by a driving-demand.

Vehicle Safety Features

Restraining systems are often implemented in fleet vehicles to improve safety systems. Restraint systems can include seat belts and airbags. Seat belts can act as active restraining systems because an effort must be made to use them. The vehicle computer system will implement a reminder system that uses a warning light to remind the driver to fasten their seat belt.

Passive restraining systems, such as airbags and automatic seat belts, operate automatically. Airbags are vehicle or occupant restraint bags that are designed to inflate quickly during a collision and deflate afterward. The airbag is typically mounted inside the steering wheel center pad and behind small panels in front and to the side of the seats. An airbag system deploys a large nylon bag during a vehicle collision. Deployment of the airbag is based on specific sensors in the vehicle. Impact sensors will measure the force of impact, and inertia sensors will detect the rapid change in acceleration. Sensors underneath the seat also detect the shift in weight. The airbag controller will analyze input from all sensors to determine if airbag deployment is required.

Fleet vehicles can be equipped with specific types of braking systems to improve safety. Conventional and the anti-lock braking system (ABS) are two examples. **Automatic braking systems** have become more common and are a safe technology that activates a vehicle's braking system when it is within a certain distance of a nearby vehicle. These systems combine video radar, video, motion sensors, and infrared signals to scan nearby objects near a vehicle; a brake control system is used to prevent a collision. Drivers may notice that the automatic braking system is used when their vehicle slows down or initiates a warning light.

Safety features, such as an ABS, help ensure driving or directional stability to reduce accidents. An **anti-lock braking system (ABS)** helps drivers maintain control of the steering wheel and can prevent trailer swing-out, skidding, and jackknifing. The ABS contains a hydraulic system like a conventional braking system but has wheel speed sensors on the front wheels and another set of sensors at the rear. These sensors can be independent of one another.

The sensors will sense when one wheel is moving faster than the other and cause the computer to send a signal to the ABS to open or close a pressure valve controlling wheel braking. During an emergency braking scenario, the ABS senses when a wheel is going to lock up and will allow a specific amount of brake force to let the wheel roll. The ABS will maintain as much braking force as needed and can repeat several times within a second. Each wheel will have a certain amount of braking force. A driver must know where the ABS lamps are located and check that they are working. The ABS lights, labeled "ABS," must come on and off briefly when the vehicle is given power. If the lamps do not illuminate or stay on, it is a violation.

Trailer sway, or **swing-out**, is when a trailer begins to swing side-to-side and causes the driver to lose control. It can occur when a tow ball or a single pivot point is located behind the vehicle's rear wheels or during a substantial change in direction. Towing under strong winds, with specific moving loads, and with uneven tire pressures can also lead to trailer sway. To prevent trailer sway, drivers should pay attention to corners that have advisory speed signs. To ensure that braking and acceleration are smooth around corners, drivers should apply a braking technique that requires the use of anticipation and scanning of the environment.

Skidding occurs when one or more tires lose traction or slip on the road, which can result in a major accident due to loss of control. During skidding, the wheels can lock, and steering is not possible. The vehicle may slide sideways or forward (plow out) while moving. When braking quickly or heavily, the ABS can reduce skidding by pulsating, or locking intermittently, so that the tires have more traction or contact on the road.

Jackknife is a vehicle motion that occurs when towing a trailer. The vehicle and trailer fold such that the angle between the vehicle and trailer/load is at an acute angle, as shown in the figure below. The motion is similar to how a pocketknife is closed.

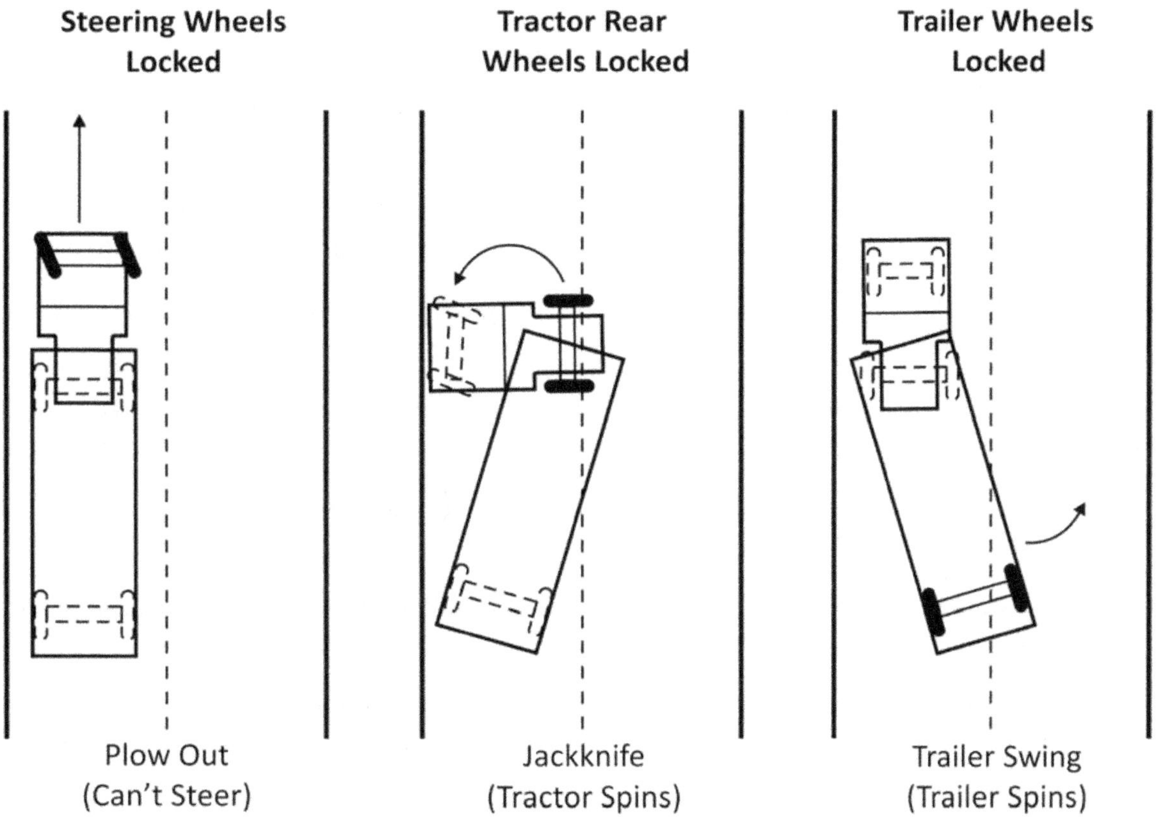

Hazard Communication and Globally Harmonized System

Hazard Communication

Hazard communication is the notification of employees concerning the noxious health effects and physical dangers of hazardous chemicals in the workplace. Workers should be clearly notified of any physical hazards (corrosion or flammability) or health hazards (skin irritation and carcinogenicity) that they will come into contact within the workplace. OSHA created the Hazard Communication Standard (HCS) to ensure that chemical information is accessible to all individuals who may interact with the substance. In addition to the HCS, all employers are required to implement a hazard communication program that encompasses training, access to material safety data sheets (MSDS), and labeling of hazardous chemical containers.

Globally Harmonized System of Classification

The **Globally Harmonized System** of Classification and Labeling of Chemicals (GHS) unifies the way countries define, classify, and communicate safety information about chemical products in the workplace. GHS always uses the same format for labels and safety data sheets. The purpose of GHS is to globally regulate hazardous chemicals and to facilitate trade, as well as to increase efficiency and reduce costs. GHS consists of two components: classification and labeling.

GHS classifies chemicals by three types of hazards: physical, health, and environmental. Chemicals in each class are assigned a category and a number or letter that corresponds to the severity of the hazard.

Health hazards pose a risk to health and include toxins, carcinogens, irritants, or corrosives that cause damage to tissue, respiratory system, or internal organs.

Physical hazards pose a risk to the physical body or the workspace. These include explosive or flammable liquids, solids, or gases, oxidizing and pyrophoric chemicals, and self-reactive substances.

Environmental hazards are chemicals that can cause damage to the environment, such as to aquatic or plant life, or the ozone layer.

GHS requires that certain information be included on the labels of all chemicals in the workplace. This information includes the class and category, hazard and precautionary statements, and a signal word.

The hazard statement is a standardized statement that describes the chemical hazard based on the class and category of the chemical.

Like the hazard statement, the precautionary statement is standardized and describes the steps to take to prevent or minimize exposure to the chemical from improper handling.

Chemical labels include signal words such as **Danger***,* **Warning**, or **Caution** to communicate the level of hazard of the chemical, based on the chemical's class and category. These signal words also appear on the Safety Data Sheet.

In addition to signal words, GHS labels include pictograms for each chemical. Framed in a red border, the pictograms visually represent the type of hazard the chemical presents.

The Safety Data Sheet (SDS) in GHS serves the same function as the MSDS for chemicals in the workplace. These data sheets include standardized information about each chemical, including the class, category, and signal word, as well as more detailed information about the hazards and precautions for safe handling and storage.

Control of Hazardous Energy

When employees work on (or service) machines and/or equipment, the appropriate energy control procedures must be used to prevent worker exposure to the release of hazardous energy. This involves de-energizing equipment, performing the correct sequential shutdown procedures, and properly detecting and isolating energy sources in conjunction with effectively using LOTO devices and equipment. In addition, the possibility of accumulating and releasing residual energy must be continually monitored throughout the service and repair process. During construction activities, common LOTO injuries and fatalities result from the hazardous energy released by concrete mixers. Isolation devices for machines and equipment include circuit breakers, in-line ball and gate valves, disconnect switches, chains, etc.

Lockout/Tagout (LOTO)

LOTO and energy isolation procedures are designed specifically for the machine or piece of equipment. The generalized lockout sequence is as follows:

1. Prepare the machine for shutdown and notify the affected personnel.
2. Shutdown the machine or equipment.
3. Identify and isolate the energy source(s).
4. Apply LOTO devices.
5. Release or relieve stored energy and monitor residual accumulation.
6. Try out the machine – for verification of procedure effectiveness.

LOTO equipment consists of padlocks and keys, locking safety hasps, blocking devices, guards, covers, and tags with warning labels stating: "Do Not Operate" (or a similar wording). Lockout materials must be in good condition and be:

- **Durable**: Materials fit for the conditions
- **Standardized**: By size and shape
- **Substantial**: To withstand stresses and prevent removal
- **Identifiable**: For visual continuity

The employer is responsible for an Energy Control Program that must include the following:

- Written energy control procedures
- Employee training
- Periodic inspection
- Enforcement and penalization with documentation
- Annual inspection

Excavation, Trenching, and Shoring

Many construction projects begin with the trenching and excavation of the site. The process is performed within varying soil types. Knowledge of soil type and classification are vital for the protection of employees who work in construction trenches. A soil classification system with guidelines and regulations (and how they pertain to soil type, trenches, and excavations) exists in **OSHA 29 CFR 1926 Subpart P**. Trench and excavation hazards include crushing, suffocation, struck by, and falls. The proper classification of any soil type pertaining to the excavation/trench, in conjunction with the application of OSHA's 1926 Subpart P, provides important information for the identification, analysis, and remediation methods that minimize these hazards.

The mechanics of a soil/trench collapse begin with the factors of soils' heavy weight per cubic foot, gravity, and compounding factors with increased trench depth and the presence of water. Stronger soils are more dense and self-supportive, while weaker soils are less dense and more granulated. Depending on soil type and saturation, the lateral pressure per foot squared increases to the point where the soil or trench can no longer support itself. These factors determine the necessity for shoring and shielding, and the requirements for angle of slope and benching techniques.

Soil Analysis and Classification

Soil analysis and classification are performed using various methods and devices. Soils are classified into four types with respective compressive strength ratings in **tons per square foot (tsf)**. These four soil types are: **stable rock, type A, type B, and type C**. When classifying soils, multiple testing methods should be performed on soil samples. Results are analyzed and recorded for determining safeguards and design techniques. Sloping and benching methods correspond to soil type. Shoring and shielding equipment are constructed of primarily aluminum and timber. Methods, terms, and devices associated with trench and excavation activities are as follows:

- **Shield**: Single unit with bracing designed to protect workers in the structure from trench collapse/cave-in.

- **Shoring**: Multiple units that protect workers by supporting the sidewalls of trenches and excavations. Shorings are designed to create a longer area of protection.

- **Stable Rock**: Solid and non-fissured rock that can be sloped vertically or 90 degrees from horizontal.

- **Type A Soil**: Cohesive soil with a tsf rating of 1.5 or greater. Designed slope can be ¾ to 1 (rise over run percentage) or up to 53 degrees from horizontal and can be benched.

- **Type B Soil**: Cohesive soil with a tsf rating between 0.5 and 1.5. Dry but determined unstable due to fissuring or being previously disturbed. Designed slope can be 1 to 1 or 45 degrees and can also be benched.

- **Type C Soil**: Cohesive soil with a tsf rating of 0.5 or less. Soil is granulated, saturated, or submerged. Designed slope can be 1.5 to 1 or no greater than 34 degrees. Type C soil cannot be benched.

- **Visual Analysis**: A visual examination method of viewing and inspecting the soil sample for cohesiveness, fissuring, layering, and granular composition.

- **Plasticity**: Measured by molding the soil sample into a small ball, then rolling it out to a minimum thickness of 1/8 inch. Cohesiveness/Non-cohesiveness is determined by holding a minimum 2-inch-long piece at one end to see if it tears apart under its own weight.

- **Dry Strength**: Determined by examining the sample behavior when crumbled by varying degrees of hand pressure. Sample reaction reveals the ratio of clay to gravel, sand, or silt.

- **Thumb Penetration**: Method of applying thumb pressure to an unconfined and cohesive soil sample. This reveals an approximation of a tsf rating applicable to soil types A, B, and C. A high degree of penetration difficulty corresponds to soil type A; a moderate degree corresponds to type B; and penetration with light pressure and minimal effort corresponds to type C.

- **Pocket Penetrometer**: A penetrometer contains a spring-loaded pin, gauge, and a stop ring on the body of the device. The pin on the end of this device is pushed into a section of a soil sample until the soil contacts the edge of the ring/stop. The penetrometer measures unconfined samples of compressive strength in tsf.

- **Tor/Shear Vane**: This method is performed on undisturbed or "in-form" soil by depressing the fins of the instrument into the soil sample until the soil contacts the backing plate behind the fins. Zero calibration ensures measurement accuracy and is done before twisting the device until the embedded portion (fins) shears soil away from the section. This instrument also measures in tsf.

The presence of water in soil and standing water exacerbate the possibility of a trench collapsing. Standing water must be removed from any trench before worker entry. In accordance with OSHA guidelines (1926.100), a registered engineer must design all trenches to be deeper than 20 feet. Site programs must consider additional hazards including: the weight of nearby equipment and associated vibration; fall hazards greater than four feet; oxygen deficiency; accumulation of toxic fumes within trenches from other equipment and processes; rescue plans; and a means of egress where the placement of ladders, ramps, etc. must accommodate workers' lateral travel to 25 feet or less.

Confined Space

Confined space (CS) activities present a multitude of hazards that can result in unnecessary fatalities. Most fatalities result from air quality issues and hazardous atmospheres. Of all confined space fatalities, the overwhelming majority have traditionally been supervisors, project managers, and would-be rescuers attempting to save a coworker or relative. Nearly 25% of fatalities are from entry into a space that already contains deadly conditions. In addition to hazardous atmospheres, confined space hazards include drowning, engulfment and suffocation, blunt force trauma, entanglement, electrocution, and lockout/tagout issues that contribute to these hazards. A confined space must have all three of the following characteristics:

1. The space must be large enough for a person to fully enter and perform work.
2. The space or area cannot be intended for continuous occupancy.
3. Any means of entry and exit must be limited.

Examples of confined spaces include tanks, manholes, vaults, trenches, and ventilation ducts. Some confined space activities require a permit before work can be performed. Permits for confined spaces require the recording of general information, hazard evaluation, initial air testing, entry preparations, communications, air test records, and the supervisor's signature. Confined space permits must be kept on file for one year. For a confined space to require a permit, it must have at least one of the following conditions:

- A structure that can trap a worker through falling walls or tapered floors, or is inward converging or tapering
- Hold or has the potential to acquire hazardous atmospheres
- Hold material that has the potential to engulf a worker
- Contain other detected hazards and lockout/tagout issues

Control Methods

Control methods also include the ventilating, purging, and inerting of hazardous atmospheres. Purging and ventilating include the use of air and/or water to flush hazardous atmospheres. Ventilating can be done via supply, exhaust, and the use of positive and negative air pressure. Inerting involves the use of nonhazardous gases, such as nitrogen and argon to push, to expel hazardous atmospheres from confined spaces.

Confined space operations must have a minimum of two workers: the **entrant** and the **attendant**. Entry supervisors authorize and oversee the operation's elements and entry. With proper training, the supervisor can serve as the attendant or authorized entrant. Supervisors must test and review the feasibility of retrieval systems (e.g., tripod/rescue harnesses for non-entry rescues) to determine system effectiveness and the potential for creating additional hazards like entrapment or entanglement.

Attendants monitor the entrant(s) and the conditions associated with confined space activities. When tracking the number of workers in a confined space, attendants must be knowledgeable of atmospheric hazards, and must monitor means, equipment, and worker symptoms while maintaining constant communication. The attendant must also monitor the activities outside of the confined space and keep unauthorized personnel away from the area. They must never perform other tasks that would interfere with these duties, and they must never enter the confined space unless relieved by another qualified attendant.

Physical Security

Obtaining a safe and secure working environment is not accomplished by simply strategizing. The staff of an organization must have adequate training to appropriately respond to diverse situations. Workplace security plans and policies address a variety of issues from a sudden crisis to an act of intentional harm. A clear understanding of security plans and policies can minimize unpredictability and panic and teach employees how to respond to a crisis.

Employees should understand security plans and how they address the physical security needs of the work environment. Workplace security plans and policies may include security measures such as control badges, keycard access systems, backup communication systems, locks on various rooms and closets, and concealed alarms. When developing workplace security plans, a team approach is vital to ensuring its success. Representatives are needed from human resources, legal counsel, security, and facilities to provide a comprehensive perspective of security needs. Once the security plans and policies are established, employees should be trained annually to review the plans and their importance.

Fall Protection

A fall from height occurs when a worker loses balance and becomes subject to the very force keeping humans on the planet: gravity. Falls from height are one of the leading causes of death in the construction industry alongside being struck by an object, caught in or between objects, and electrical hazards. **Fall hazards** are worksite conditions or materials that may cause a fall and include falls from ladders, scaffolding, structural steel, and roofs. They result from unprotected sides on raised platforms such as roofs, scaffolding, or uncovered floor or roof openings (others may include unsecured ladders or scaffolding in poor condition). **Same-level falls** are generally caused by trip hazards and poor worksite housekeeping that result in neglected ground obstacles or impediments (one of the leading causes of injury in the construction industry). Fall protection measures should be taken to prevent injuries.

Fall Hazard Sign

Duty to Provide Fall Protection

OSHA's requirements regarding fall protection can be found in 1926, subpart M. For most working surfaces in construction, such as roofs, edges, and floor openings, fall protection must be used six or more feet above the ground or the next lower working surface. Fall protection is also required when a fall is possible onto dangerous equipment, machines, or materials when less than six feet in height.

OSHA's general rule to provide fall protection at 6 feet differs in some situations in construction. Fall protection is required on scaffolds 10 feet above the ground or the next lower working surface. Scaffold safety provisions can be found in 1926, subpart L. Fall protection on aerial lifts is always required, regardless of height. Although treated differently than scaffolds, OSHA's aerial lift requirements can also be found in 1926, subpart L. In some circumstances involving steel erection, fall protection must be provided 15 feet or more above the ground or the next lower working surface (20 feet in some scenarios.) These requirements can be found in 1926, subpart R. Fall protection guidelines for stairways and ladders can be found in 1926, subpart X.

Fall Restraint vs. Fall Arrest

The majority of fall protection systems in construction can be categorized as either fall restraint or fall arrest. Fall restraint includes guardrails and also harness devices that will not allow a fall from height to occur. With a fall arrest system, such as a **personal fall arrest system (PFAS)** or safety nets, a fall can happen, but a device stops the fall.

While contractors are expected to select the system that provides the most reliable means of protecting employees from fall hazards, OSHA generally prefers restraint systems to arrest systems because a fall does not occur at all when a restraint system is installed, such as a standard guardrail system. Fall restraint systems also tend to be less complex than arrest systems and so do not require the level of advanced technical proficiency needed to effectively employ a fall arrest system.

Guardrails

Guardrails are a fall restraint system that is most frequently installed on flat working surfaces. Although technically allowed in some circumstances, the use of guardrails on sloped surfaces is not advised for various reasons, most importantly because, even when strong enough to prevent a falling employee from breaking through, the employee could be hurt when contacting the guardrails.

OSHA's requirements for guardrails in construction, found in 1926, subpart M, specify that the top rail be installed at 42 inches from the working surface, plus or minus 3 inches; so, anything between 39 and 45 inches is acceptable.

When a midrail is installed, it must be approximately half the height of the top rail. Midrails are not mandatory but are very common. An alternative to a midrail is to install a wire mesh or other suitably strong material from the top rail down to the working surface. Vertical supports (balusters) may also be used instead of a midrail but must not be farther than 19 inches apart.

A guardrail system must be strong enough to stop a fall; OSHA specifies that the top rail be able to withstand a force of at least 200 pounds (890 Newtons). While caution tape may allow some notification of the proximity of a roof edge on account of its highly visible color, it should never be permitted to serve as a guardrail system because it will not stop a fall.

Machine Guarding

Machine guarding refers to safeguards that are implemented or attached on machines that prevent harm to persons and property. Any aspect of a machine has the potential to cause harm. When a machine causes harm to operators or bystanders, the hazard must be addressed to prevent further harm. OSHA has determined machine guarding hazards within specific industries. Additionally, it creates safety standards and materials related to machine guarding in a specific industry as a resource. An effective tool of machine guarding is being able to recognize a hazard and incompetent machine guarding.

According to OSHA, all machines consist of: (1) point of operation; (2) power transmission device(s); and (3) operating controls. Point of operation is where the machine executes its designed purpose. Power transmission is the portion of the machine that transmits energy so that it may operate its designed purpose. It may include such things as gears and belts. Operating controls are mechanical and electrical controls that allow the operator to manipulate the machine's power as desired – such as an off and on switch. While machines may be similar under this model, safeguard needs vary depending on the characteristics of the machine, operator, and environment. Varying parts to consider include, but are not limited to, nip points, flying chips or sparks, and revolving machine parts.

Powered Industrial Vehicles

Trucks

Trucks are large, motorized vehicles that may be used to transport materials or goods. **The Department of Labor (DOL)** lists truck driving as one of the most dangerous occupations within the United States. Adhering to safety regulations and procedures increases protections for the driver, property, and other drivers on the roads. Additionally, it limits liability that may occur from resulting accidents.

Truck drivers should engage in defensive driving practices. Using defensive practices will decrease the likelihood of an accident. These practices include monitoring blind spots, emergency preparation and training, using turning signals, and traveling within speed limits. Preventative maintenance schedules should be developed to ensure that the truck is operating in a sufficiently safe manner. Safety inspections should be conducted pre-trip and post-trip to assess whether there are any potential issues that may occur during operation. Additionally, drivers should be trained to avoid distractions, such as eating and operating cellular devices during operation. Sufficient rest and break periods should also be utilized to avoid fatigue that may affect driver distraction and reaction.

Drivers should have access to sufficient tools to respond to hazardous road conditions. GPS tracking is a tool that allows a supervisor to monitor the truck and re-route it if necessary to avoid hazardous road conditions.

Forklifts

Forklifts, also called lift trucks, are powered industrial trucks that various industries use to transport materials with efficiency and ease. The forklift may lift large items or heavy items in bulk and relocate the items to a desired location. Forklift and truck hazards vary depending on the truck, environment, and operator. For example, a forklift is more likely to be in a tipping accident because of its ability to pick up heavy objects and lift them – which may cause an uneven distribution of weight. Environments also present potential hazards, such as retail stores having more pedestrians in a surrounding area than an industrial site. Before a person may operate a forklift, it is important that training must be completed to ensure that the operator is competent. Subsequently, operating a forklift must be done under a qualified supervision.

Operating a forklift safely requires that the operator be prepared for potential hazards and be aware of the surroundings and forklift controls. Before use, it is best to inspect the forklift and address any maintenance concerns that may arise. This ensures that it operates properly and increases safety. Utilizing common safety operating practices decreases the likelihood of an accident occurring. Finally, when moving objects, it is best to determine the potential hazards of the load and how to effectively avoid those hazards while operating the forklift. A substantial factor in this is to be aware of the load composition and the effects it may have. If a forklift needs to load or unload materials into a semi-trailer, the semi-trailer must be evaluated for potential safety concerns.

Cranes

Overhead cranes are used to hoist and transport heavy loads on construction sites. Like the rigging and hoisting process, crane operation presents hazards from improperly secured materials slipping, falling, or swinging out of control. Crane operation also presents electrical hazards and hazards associated with overloading the equipment.

During crane operations, heavy loads are hoisted over a work area where workers are usually present. If the loads are not properly secured during rigging, they can slip, and materials can strike workers or cause property damage. Workers can be pinched, crushed, or caught between swinging loads if crane operators are visually impaired or incompetent, or if the operators fail to pay attention to their work.

In addition to risk of shock from electrical tools, parts of the crane such as the boom can come into contact with active power lines and cause injury or death to the operator or to workers in the vicinity.

Exceeding the operational capacity of a crane can result in structural damage to the equipment, which can result in malfunctions that can cause serious injury. If the crane is overloaded, the load might slip and fall, causing injury to workers below.

To prevent loads from slipping and crashing to the ground, workers should be trained to properly secure loads and remain within the operational capacity of the equipment. To prevent electrical shocks, crane operators should be aware of power lines within 10 feet of operations. Workers should always wear protective equipment on the work site.

Scaffolding

Scaffolding is a work platform that can be supported from the ground or suspended. This equipment is used to support workers and associated materials during construction activities. The OSHA provides standards for 25 types of scaffolding. Ladder jacks, pump jacks, aerial platforms, and aerial lifts are also considered scaffolds. For equipment to be recognized as a scaffold, it must be:

- Temporary
- Elevated
- Designed to safely support both workers and material

The most common type of scaffold is **tubular welded frame**, which is assembled from prefabricated framing and cross braces. Walking and working areas should be equipped with scaffold-grade lumber.

Decks must be fully planked where work is performed, while areas exclusively used for walking and traversing require a minimum planking width of 18 inches. Scaffold parts must be of the same type with no mixing of components. Supported systems must have a load capacity of four times the intended load, while suspended systems must support six times the intended load. Scaffolds require fall protection beginning at a height of 10 feet without the presence of guardrails. As a general rule, scaffold units exceeding a height-to-width ratio greater than 4:1 must be tied in. Tie-ins are also required at width intervals of 30 feet. Scaffolds should never be climbed upon and they require safe points of access with the presence of ladders. Supported units on asphalt or soft ground require baseplates coupled with mudsills to prevent sinking, twisting, or potential collapse. A competent person using a tagging system should inspect scaffolds before each shift.

Scissor Lifts

Scissor lifts are commonly encountered in construction. Most scissor lifts in use today are a type of mobile scaffold; scissor lifts that require movement and placement to be accomplished with another device are not common in the construction industry, although they do exist and have their purpose. If a scissor lift is mobile, then OSHA treats it as a mobile scaffold, and applicable safety requirements for scaffolds will apply.

Most of the general safety principles that apply to fixed scaffolds will also apply to scissor lifts, for example, strength and suitability, proper footing, sufficient working space on platforms, and the common use of guardrails. As with nonmobile scaffolds, a PFAS is not mandatory when working on an elevated scissor lift platform. Some scissor lifts do have an anchorage point on the deck for attaching a PFAS, but this is not very common, since PFAS is not mandatory as is the case with aerial lifts. A guardrail system is the predominant means of fall protection employed on most scissor lifts.

Additional precautions are necessary for powered and mobile scissor lifts. When moving a scissor lift, employees should be permitted to remain on the platform only when it will be moved slowly and safely, when the traveled surface is within three degrees of level, and when there are no holes or obstructions that could prevent safe movement. The scissor lift should be lowered when traveling to reduce tipping. OSHA requires the scissor lift to be no higher than twice the width of the base when traveling with mounted employees; the only exception is when the device's engineering parameters are such that it will remain stable when traveling with a height to base width ratio exceeding 2:1.

Mobile scissor lifts are sometimes misidentified as aerial lifts. While both types of equipment involve elevated working platforms, the methods of fall protection permitted by OSHA's 1926 standards are significantly different. An aerial lift, as defined by OSHA, is easily identified by a boom that is extensible (telescoping) and/or articulating (multiple planes of movement). A scissor lift will not have any components that telescope or articulate; it will only go up and down on its scissor mechanism.

Aerial Lifts

An ASP should understand that, while OSHA's guidelines for aerial lifts can be found in a section contained within 1926, subpart L, scaffolds and aerial lifts are not all treated as scaffolds, and general scaffold fall protection requirements do not apply to aerial lifts.

Unlike scissor lifts and other scaffolds where guardrails are the norm, OSHA requires that a harness be used whenever working from an aerial lift, even when guardrails are installed. The requirement to wear a harness applies regardless of the height of the platform, even when the platform is less than six feet from the ground. A PFAS utilizing a full-body harness must be used when a fall from the aerial lift platform is possible; this is the same treatment as when working on any other surface when guardrails are not feasible.

An ASP should advise that the PFAS be attached to an anchorage point on the deck, and not the boom, railing, or anything else higher up whenever possible. If this is done, then the total fall distance prior to the arrest will be minimized in the event of a fall from the platform. OSHA allows up to a 6-foot drop in a PFAS, although this should always be minimized, since an arrested fall can still cause back or other injuries.

Assume the total length of the PFAS components (not just the lanyard) will arrest a fall at six feet exactly. If the PFAS is attached to the boom or the top of the guardrail or bucket, then the fall arrest distance will be approximately six feet. However, if the anchorage is on the deck, then the fall arrest distance is reduced because the fall will be limited to the length of material that goes beyond the top of the platform and not its total length. Minimizing the fall arrest distance will also help to minimize the likelihood of an injury resulting from an arrested fall.

If the anchorage point and chosen PFAS components are such that an employee cannot physically go over the edge, then there is no fall arrest situation at all. In such situations, OSHA even allows the use of body belts, and not a full-body harness since there is no fall potential. OSHA permits the use of a **personal fall restraint system (PFRS)** utilizing a body belt when a fall is absolutely not possible, although this practice has mostly become unpopular on aerial lifts, and an ASP should advise the use of a full-body harness even when a restraint system would be in compliance with OSHA's standards. Another consideration when introducing body belts for use solely in restraint situations is that employees may see them used and incorrectly assume that they are acceptable in fall arrest situations and utilize them as such.

There is one instance when not wearing a harness in an aerial lift is allowed and may well be advisable. This is when working over water and the drowning hazard is greater than the fall hazard. When drowning is deemed the greater hazard, an ASP should ensure that employees are wearing an approved personal flotation device and that means are available to get them out of the water as soon as possible such as a small boat. When correcting hazards in construction, an ASP must avoid the introduction of even greater hazards in the process.

Using Hazard Identification Methods

In an effort to implement a national standard occupational safety and health program, the U.S. Department of Labor established the OSHA of 1970. Prior to OSHA, states established and managed their own occupational safety and health programs. These programs varied from state to state, with some more strict and some less strict in their monitoring and enforcement of safety programs. While some employers neglected to establish a safety program or ran a poorly managed one, other employers voluntarily invested significant resources to establish and maintain an occupational safety program.

In order to equalize the perceived economic advantage held by nonparticipant business owners and corporations, the government created OSHA to establish national minimum occupational safety standards for all business entities to share equal accountability for the safety of their employees (although states still typically run their own workers' compensation programs and OSHA departments). OSHA operates in concert with other regulatory agencies. The OSHA is responsible for the establishment and enforcement of mandatory health standards. The National Institute for Occupational Safety and

Health (NIOSH) is the research arm of OSHA and enjoys the same workplace entry privileges as OSHA. The Occupational Safety and Health Review Commission (OSHRC) is the resolution arm of OSHA and will attempt to resolve citation disputes.

OSHA has a right to conduct unannounced site inspections but will do this with management consent or warrant. Site inspections are the primary method of enforcement utilized by the agency.

The five inspection methods include general schedule, complaint, fatality/catastrophe, imminent danger, and follow-up. General schedule inspections are conducted in high-hazard environments. Complaint inspections are generated internally and are filed confidentially by the employee or employees' representatives. Fatality/catastrophe inspections are conducted when a fatality occurs or when three or more personnel are hospitalized. Imminent danger inspections are instigated by any individual or party declaring such a scenario. Follow-up inspections tend to be associated with a past infraction or citation, and the subsequent inspection serves to verify infractions or cited items have been effectively mitigated.

The implementation of a worksite safety program must necessarily revolve around the baseline or initial site risk assessment whose main purpose is to build and document a base of reference for known and potential hazards unique to the worksite.

This initial audit provides the foundation for an active and continuous hazard identification program that includes daily safety walks (DSWs) and scheduled or impromptu inspections. Worker participation, feedback, and "buy in to the worksite" safety programs will go a long way toward perpetuating a sound safety culture.

The steps following hazard identification include the control measures applied to assess and mitigate such hazards. As stated earlier, the hierarchy of control measures includes elimination, substitution, engineering, administrative, and PPE. There is a possibility that a control measure for a hazard may be changed, should a variable present itself in circumstances, location, equipment, or materials. Documented evidence must reflect any worksite changes using a management of change (MOC) procedure.

The hazard identification program will be an ongoing process that manifests itself in monthly inspections, weekly inspections, and DSWs. As important as scheduled inspections and worksite monitoring for hazards are, nurturing an atmosphere of trust and common interest by the ASP in the workplace is of equal importance, as this encourages personnel to participate and provide hazard identification feedback.

The success of a worksite safety, health, and environmental program depends on some of the following factors:

- The installment and maintenance of a post-comprehensive hazard survey identification program
- Occupational safety and health surveys—worksite surveys include hazard identification, material inventory including hazardous chemicals (MSDS), hazard communications and notices, air sample analysis, noise-level surveys, and ergonomic surveys
- The ASP must be aware of changes to materials, equipment, methods, and simultaneous operations that divert from the normal standard operating procedures (SOP) in place. The ASP must reconcile any potential MOC scenario with the requisite hazard analysis and mitigation

measures including the documentary evidence relative to the action. This includes CAPA plans. All hazards must have related CAPA plans with each identifying root cause and a mitigating solution or solutions assigned to personnel with a required deadline for plan implementation. All CAPA plans must be approved by project management in accordance with regulatory bodies and regulations. Any CAPA not corrected will put the project at risk for further incidents and noncompliance.

Assessing and Analyzing Risks

The employer is responsible for hazard detection, hazard assessment, and protecting employees from workplace hazards. The process is primarily performed through a basic three-step premise: **Recognize, Evaluate, and Control**. This may seem simple, but each step in the process requires a multitude of factors, concepts, and considerations. Control methods should be considered based on those with the greatest effect on the hazards being analyzed. In turn, this produces the most effective impact with respect to relevant outcomes.

A major aspect in establishing a hierarchy of controls is understanding that the result of corrective measures is the achievement of an acceptable risk level. This means that the probability of occurrence, exposure level, and degree of harm are as low as reasonably tolerable and achievable for the situation. Factors including feasibility, cost with respect to expected outcomes, and the potential to create new or additional hazards can impact the means and level for addressing a hazard. Such situations can result in a higher degree of residual risk where reassessment and additional measures may be required. These extenuating elements express a behavioral flow and complications that can arise when addressing hazards and risk. Such factors require a systematic way of thinking when applying a hierarchy of controls.

Throughout the hierarchy, an important aspect to note is that higher controls reflect more effective strategies that are both proactive and preventive, thereby eliminating or greatly reducing risk. These controls rely on safer designs, reengineering equipment, and replacing material elements where corrective actions are directed at hazards inherent in a system. Lower hierarchal controls reflect strategies that are more reactive with respect to incident occurrence. They result in higher expenses that are compounded by incident costs. In contrast to design and engineering measures, these controls are geared toward modifying human behaviors in the presence of hazard risk.

When establishing controls to eliminate or reduce risk, all elements of a higher order should be considered and exhausted before selecting controls from lower in the hierarchy. This presents a more thorough analysis of controls and results in more effective measures.

Practitioners of safety can evaluate hazard risk via risk matrices. Risk matrices function by determining the potential of event frequency (or likelihood of occurrence) versus event severity level to determine hazard risk. This process can be used in conjunction with JSAs and other analysis tools. Risk matrices are derived of alpha (qualitative) or numeric (quantitative) elements to yield an overall risk level or score. Subsequent to determining risk level, appropriate corrective actions are applied to eliminate or reduce risk.

The **Z-10 Risk Assessment Matrix** and other risk matrices from the **American National Standards Institute (ANSI)** provide hazard risk prioritization (for both frequency and severity) from which risk level can be determined. Using this strategy, the rate of incident frequency (or likelihood of occurrence) can be plotted versus the incident severity level to determine the overall level of risk. This and other matrix types can derive **alpha** (qualitative) or **numeric** (quantitative) elements from which a risk level can be attained. Subsequent to determining risk level, an organization should consider mitigation measures to eliminate or reduce the risk. Control measures should be prioritized through a hierarchy and feasibility, beginning with elimination, substitution, and engineering controls. When a hazard is eliminated, the risk is eliminated as well. Efforts that result in reduced risk require additional assessment for the possibility of single or multiple controls. Lower hierarchal controls include warnings, administrative controls, and personal protective equipment (PPE).

Here's a look at two of these matrices:

Severity / Probability	Catastrophic (1)	Critical (2)	Marginal (3)	Negligible (4)
Frequent (A)	High	High	Serious	Medium
Probable (B)	High	High	Serious	Medium
Occasional (C)	High	Serious	Medium	Low
Remote (D)	Serious	Medium	Medium	Low
Improbable (E)	Medium	Medium	Medium	Low

Likelihood / Impact	Rare (1)	Unlikely (2)	Possible (3)	Likely (4)	Almost certain (5)
Catastrophic (5)	5	10	15	20	25
Major (4)	4	8	12	16	20
Moderate (3)	3	6	9	12	15
Minor (2)	2	4	6	8	10
Insignificant (1)	1	2	3	4	5

Providing Financial Justification of Hazard Controls

Hazard controls refer to programs that involve necessary requirements to protect employees from potential hazards. These programs should be written and monitored to determine whether current practices are effective. This will lead to increased safety and decreased likelihood of harm to persons and property. An effective hazard control program should be one that sufficiently trains employees, determines appropriate controls, utilizes hazard control regulations and guidelines, and objectively measures conduct and compliance.

Determining the proper hazard control(s) based on financial feasibility, may be completed by a gap analysis. First, hazards should be identified so that there is a specific matter to be addressed. This may be completed by conducting employee interviews and reviewing relevant documents, such as incident reports. Interviews will provide input from employees operating under the relevant circumstances and relay common experiences. Reports will also show factual assertions, including the frequency of a hazard.

Next, control options should be identified. The goal should be to immediately eliminate or limit substantial hazards. Options that may give rise to new hazards should be discarded because it will increase cost. Two categories will then be created regarding potential options. The first category will be the cost of implementing a proposed option. The second category that will be created is the benefit of implementing the corresponding option. There may be multiple sub-costs and sub-benefits, depending on the number of options. Ensure that each cost and benefit is assigned a monetary value.

After all costs and benefits have been identified and valued, all values should be combined to show a summed value. Depending on how the analysis is conducted, there may be multiple cost and benefit values, or there may be an aggregated overall value. The difference between the two is that individual options may be evaluated, or options within a total program may be evaluated. By evaluating individual options within a program, it will show which options may be financially feasible and which may be discarded.

Completing each individual analysis or aggregated analysis will show an objective monetary value that can be applied to each option or to the program as a whole. Conducting this analysis process should be based on financial considerations and will show options that have more benefit than cost. Providing more benefit gives an objective justification on what control options should be utilized by the company. This is a calculated and reasoned conclusion and gives an evidentiary basis to financial considerations. It is imperative that each value be based on objective data, whether it be previously available data or current information. Compiling all the relevant reports and information provides objective justification and is not speculative. Making financial decisions on speculative information and matters increases risk. Speculative matters should not be included in the gap analysis because it introduces error and misconception, and it will undermine the credibility of the proposed financial justification.

Implementing Hazard Controls

Site-Specific Safety Plans

For any new work project, particularly involving building, management planning encompasses tasks such as:

- Identifying the necessary materials and equipment for the project
- Assessing the appropriate number and type of workers needed to complete the project
- Creating the project schedule
- Tallying the associated costs

For good safety performance throughout the project's life cycle, project planners apply risk assessment principles to detect losses associated with compromised workers, public safety, and materials. These elements are included in the planning stage.

Effective safety planning requires owners to leave work activities in the hands of contractors who can more effectively orchestrate project safety. This reduces ownership liability, demonstrates top-down leadership, and places a fundamental focus on safety through design.

Planning begins with the cultural assessment at all levels, including program and project managers, contractors and subcontractors, and all associated workers. From a safety standpoint, the main goal is to maintain an injury-free workplace through a culture that pursues safe work practices and emphasizes both individual and team responsibility. The following illustrates some of the many essential elements for effective site-specific safety planning.

Work Means, Methods, Equipment, and Materials

To ensure effective safety performance, organizations must fully comprehend the entire scope of work activities that will occur during the project. This understanding is critical for site safety preplanning. Knowledge of activities and processes will aid the participants in anticipating the potential system deficiencies that will require improvements as well as the hazards that can be eliminated or mitigated before the project begins. This is vital to ensure project safety processes align with other components such as material selection, scheduling, and the work to be done.

As part of preplanning and follow-up, a safety practitioner must question components involving performance and compliance for each job process that will comprise the project. This questioning process can be designed as an abilities assessment form to obtain "yes" or "no" answers or as a checklist for rating performance criteria. These assessments or checklists help determine what improvements should be made before work begins. The following outlines some of the many work processes (methods, equipment, materials), with examples of respective performance/compliance elements.

Fall Protection

- Does the design reduce the number of stairs and ladders needed?
- Has fall-prevention equipment been considered before fall-protection devices?
- Is the system designed with or to receive anchor points and fall-restraint devices?
- Were considerations given for providing fixed ladders and work platforms for work at heights?

Lockout/Tagout—Hazardous Energy Control

- Has controlling the release of hazardous energy been considered as part of the design?
- Do lockout/tagout (LOTO) procedures exist?
- Have the prospective energy forms been identified?
- Are LOTO devices accessible, standardized, and in the required condition?

Electrical Safety

- Does the electrical system meet all applicable code requirements?
- Have grounding procedures for tools and equipment been considered?
- Is high-voltage equipment isolated as required?
- Are components adequate where combustible vapors may exist?

Walking and Working Surfaces

- Will walking surfaces be designed of or designed to incorporate nonslip materials?
- Does the design allow for walkways free of obstruction?
- Do all floor and wall openings meet OSHA requirements?
- Will utilities be directed away from walking surfaces?

Confined Spaces

- Has confined-space work been eliminated or reduced in project design?
- Have items deemed as confined spaces been assessed for permit requirements?
- Has the potential for introduced hazards been assessed with confined-space work?
- Have monitoring for hazardous atmospheres and other confined-space safety requirements been considered?

Segregation and Separation

Segregation and separation refer to engineering controls designed to ensure that hazardous materials are handled safely. **Segregation** involves keeping materials secure in specified areas with restricted access, so physical security is a key component of ensuring proper segregation. Guards, barricades, secure room entry systems, and locked cabinets are often used to restrict access and segregate materials, particularly when the materials are hazardous, controlled substances, and/or valuable. Other methods of segregation seek to contain materials, such as using ventilated, enclosure, and/or sealed systems to control dust. Likewise, sticky mats, anti-static devices, and water spray systems can be used to segregate particulates. Glove box and glove bag systems are often used to isolate and contain powders when mixing, drying, or otherwise manipulating the material. Aside from security and containment measures, some methods of segregation also involve keeping materials under certain conditions, such as refrigerators, freezers, and cold rooms.

Separation involves keeping materials isolated to minimize the threats they pose, especially in combination with one another. The first step is to properly label all the materials and then to separate the materials appropriately. All materials must be further segregated based on their compatibility to avoid the materials from being in close proximity. For example, chemicals should be separated based on whether they're organic or inorganic, and then they should be segregated according to their properties and compatibility. When separating materials, safety professionals must be cognizant of the following properties: corrosivity, flammability, reactivity, and toxicity.

Corrosive materials should be separated based on whether they are an acid or base, and the different classes should be segregated in a cool and dry ventilated area. Flammable materials should be kept in a flammable storage cabinet or an approved refrigeration system; both of which must be labeled with a warning that they are flammable and that fire must be kept away. Reactive materials must be separated and then segregated based on whether they are oxidizing, peroxide forming, pyrophoric, or water reactive. Toxic materials need to be separated and secure, and then the isolated area must be specifically labeled with the materials' class of toxicity, such as carcinogen or reproductive toxicity. Additionally, toxic materials might require a secondary containment system to localize and minimalize the risk of accidents.

Monitoring and Reevaluating Hazard Controls

Historically, the construction industry sector has had a highly disproportionate number of workforce injuries and illnesses when compared to other industries. Therefore, policy and plan criteria for safety must go beyond compliance as a minimum. Many construction accidents stem from an organization's ineffective structure and the absence of a safety culture. This includes the lack of coordination and communication, as well as deficiencies in detecting existing hazards. These deficiencies result in diminished safety levels, poor performance, and more loss incidents. Hazard awareness and identification are the keys to eliminating or reducing risk. Effective hazard recognition requires training and should be performed by qualified personnel.

Monitoring Hazards

A **hazard** is any unsafe act or condition that, if left unchecked, can result in injury and/or illness. Hazards can also negatively affect people, property, equipment, and the environment. Identifying hazards is a vital part of maintaining a safe work environment, and hazard recognition should not fall solely on the shoulders of safety personnel. Hazard recognition, assessment, and mitigation strategies are most effective when there's a strong program and all layers of the workforce are committed, including project or site ownership, management, supervisors, maintenance, and laborers. Hazard assessment and risk reduction efforts can be proactive, in situ, or reactive.

Failure Mode and Effects Analysis (FMEA) is a proactive tool for hazard recognition and mitigation strategies. This process is used in the engineering and design phase to identify and describe potential system failures before an incident occurs. FMEA provides safety professionals with useful information that can help reduce risk in a process. A safety professional compares and contrasts pertinent acquired data with historical data and relevant information to identify and examine potential failures. Mathematical modeling can also analyze reliability and failure rates.

Hazards can be introduced in the process planning stages, and then magnified when different processes come together on a job site. Feasibility, cost, and lack of coordination between entities are major contributors that can introduce hazardous conditions into design processes. Eliminating hazards during a project's design stage is an effective mitigation strategy because it is proactive. It can reduce or eliminate an array of unwanted conditions that would otherwise be overlooked. Built-in hazards have three modes, which are described as follows:

1. **Dormant**: A hazard created by the inclusion or omission of elements. It lies undetected for subsequent arrival.

2. **Armed**: A hazard that's triggered or loaded through the combining or assembling of separate processes.

3. **Active**: A situation where a combination of elements actuates dormant and armed hazards into events.

Hazard identification, assessment, and mitigation is proactively critical to work safety compliance. Such review of potential workplace illnesses, injuries, and incidents is required by law. The ASP will assist in worksite efforts to foster an environment of participation-based hazard identification. The success or failure of a project rests on a successful hazard identification program and the documentation that validates it.

Hazards may include, but are not limited to, potential worker threats. First, the workplace must be examined for existing threats. These should be thoroughly documented and mitigated whenever possible. Any and all existing equipment manuals, chemical safety data sheets (SDS), OSHA logs, documentation on previous injuries, worker injury/illness patterns, past safety committee meeting minutes, or any other pertinent documentation must be reviewed prior to project implementation. Neglecting to do so can result in the same hazards, all of which may be avoidable.

After reviewing past hazard documentation, a current inspection of the worksite is essential in identifying and mitigating potential new threats. All equipment maintenance, as well as electrical and chemical hazards, must be assessed. Emergency procedures must be reviewed for accuracy in accordance with the existing workplace. Workplace organization and flow should be examined for current threats as well.

Any and all health hazards must be fully noted and documented. All chemical, electrical, physical, biological, and ergonomic hazards and past incidents should be assessed. Important tools may involve SDS and past worker medical records. Other tools such as air sampling may be required according to the planned project.

Lastly, hazard identification, assessment, and mitigation strategies should include full incident investigations. A process of reporting, documenting, identifying the root cause, and having corrective action plans in place must be part of hazard strategies. It is wise to have both emergency and non-emergency hazard workflows in place identifying personnel responsible, reporting and communication structures, documentation guidelines, incident templates, and corrective/preventive action requirements.

The goal should be to reduce risk and decrease hazards. Some hazard studies indicate risk reduction remains the focal point of project HSE planning, although further studies suggest that decreasing the hazards involved can lower the risk of individual tasks. Hazard identification is a critical tool and is utilized by the ASP to qualify and quantify the most practicable worksite hazard and risk reduction measures.

Hazard assessments must contain clearly identified hazards and mitigation plans for each. Mitigation defines the measures taken and resources allocated toward worksite risk reduction of assessed hazards; each hazard is broken down, analyzed, assessed for its unique risks, and has the proper mitigating measures applied and documented. Failure to conduct hazard identification, assessment, and mitigation will lead to undesirable, unforeseen, and reckless work.

Following up on Corrective Actions

As part of design safety planning, corrective actions are developed to address hazards in a proactive manner (before work activities begin). The main objective of design safety is hazard prevention through developing controls that eliminate risk. Risks that cannot be completely eliminated must be reduced to a level as low as reasonably practicable. Controls are developed as a result of hazard detection where risk analysis, prioritization, and a hierarchy of controls were guiding factors in decision-making and intended outcomes. This means that the corrective action was developed with an expectation of performance and effectiveness in mind.

Corrective actions, whether pre- or post-design, often require performing cost-benefit analyses that consider the input of money and level of effort and theorize the point in time where a return on investment (or monetary balance) may be achieved. The question, "Are/were the results of the action worth the time, money, and effort?" further necessitates evaluating control performance and efficiency. Follow-up efforts must be conducted to evaluate the effectiveness of corrective actions.

Evaluating performance through follow-up efforts measures the effectiveness of corrective actions against expectations, or the action's planned level of performance. Results indicate whether or not the corrective actions sufficiently addressed the target hazard and reveal the amount of residual risk. Corrective actions must be analyzed with great caution prior to implementation to avoid introducing a new or additional hazard into the project design.

Follow-up procedures can be conducted as follows:

- Inspect/review performance of new equipment and materials.
- Assess operational conflicts such as scheduling and interference.
- Review design/layout modifications to reduce ergonomic and fall hazards.
- Incorporate as part of change management (e.g., modified or new equipment, materials, processes, projects).

Following up on corrective actions is vital for effectively managing process safety and risk and continually improving safety for workers, the public, and the environment. Follow-up can result in modifications to worker behaviors and prescribed procedures, improvements to program-level elements, as well as have a positive impact on organizational approaches regarding design safety with future projects. Effective corrective actions and follow-up rely on the combined efforts of employers, the safety professional, design engineers, and other key personnel. Follow-up and other risk management methods should be reviewed and adjusted on a regular basis to ensure continuing efficiency.

Conducting Incident Investigations

Following an incident, the first fact-finding efforts will range from a generalized survey of the situation to a formal investigation. The formality and complexity of the action taken will generally depend on the severity of the incident; minor near-miss incidents will not require a formal investigation involving multiple parties, and a significant injury investigation must not be limited to a brief survey.

Whichever investigation technique is employed, it must be adequate to determine probable causes; the objective of any incident investigation is to prevent a similar incident from occurring. The technique must also be sufficient to satisfy the program and policy requirements of the contractor, hiring client, and any other relevant third parties. There may be multiple parties requiring information, and there will be times when their requirements and agendas are not universally mutual. An ASP will be a source of guidance during an incident investigation and must be sufficiently knowledgeable of the organizations involved in order to effectively communicate with them during the investigation.

Since the incident has already happened and an ASP cannot travel back in time to prevent it from happening, an investigation will always be a backward-looking effort employing deductive analysis to identify all factors that could have contributed to the event.

When beginning an incident investigation, the priorities are to arrive safely, size up the situation, care for the injured, and lastly, protect property.

The most immediate supervisor, such as a foreman or project manager, is the lead person in most routine incident investigations. Since a foreman can be expected to know the work area and personnel better than anyone, it is reasonable to assume that the foreman will also be the most appropriate party to identify the causal factors that may have led to the incident. For this reason, it is not advised that an ASP or safety administrator be the lead person in routine incident investigations, and also because the supervisor who exercises control over the operation should ultimately be the person responsible for correcting the causal factors in the incident.

During a routine incident investigation, the standard role of a safety manager is to assist the supervisor conducting the investigation through observation and technical guidance as required. However, the supervisor must be effective in this role; otherwise, the fact-finding effort may result in useless data. In such a situation, the ASP or safety manager will need to have a more hands-on role in the investigation; depending on the supervisor's deficiencies, the ASP may even need to conduct the investigation.

For incidents involving fatalities, very severe or multiple-injury scenarios, and very costly property damage, it is advised that senior management formally conduct the interview. These scenarios may also involve lawyers and other third parties such as liability and insurance professionals. For these types of high-gravity events, the ASP will frequently serve as a technical advisor during the course of the investigation.

Root Causes and Causal Factors

Management deficiencies are unifying characteristics of root causes. Deficient management practices ultimately empower personnel to introduce the causal factors that make an incident possible, and there are frequently multiple causal factors in an incident. For example, a particular equipment malfunction may be identified as a causal factor in an incident that injured an employee, and the decision of senior management to purchase inferior and unsafe equipment may be identified as a root cause from which the causal factor is derived. Or perhaps the contractor's human resources department hired personnel who were unsuitable for their roles, and this practice resulted in the introduction of causal factors in the incident, such as employee errors, inadequate job and safety training, and other human factors.

When attempting to identify cause-and-effect relationships for hazard mitigation, the cause can be misidentified or found to be an undesirable event with yet another cause. *Root cause analysis (RCA)* is a technique where the safety professional describes the negative effect or loss to define the incident cause. This technique requires complete evaluation of the event and the hazards inherent in the system, determining deviations from acceptable requirements and analyzing situations essential for the event to occur. To define an incident, a safety professional determines the root cause through a series of who, what, where, when, and how questions. This process can reveal a myriad of undesirable conditions that can be mitigated.

Actions such as corrective/preventive measures, disciplinary proceedings, and informational meetings with third parties like safety committees or union representatives should only take place once useful factual data has been gathered; actions taken without factual data are counterproductive.

Although the facts may ultimately reveal that disciplinary measures are necessary, the initial analysis must be fact-based and not oriented toward finding fault for the purpose of determining appropriate disciplinary measures. A proper investigation identifies the root causes of the incident through fact-finding, not fault-finding.

Employee interviews should take place as soon as possible because important details may be forgotten, and the general clarity of information recall will degrade quickly in the first few days following an incident.

The best place to interview a witness is the incident site. The witness will be able to identify points of reference on the site and more fully explain the details involved. Moreover, communication errors are less likely to occur because the interviewer will not have to presume what the witness is saying about a specific site detail when both are present.

The incident site will not be an appropriate interview venue when the incident is very bloody or particularly ghastly in any regard because many employees will be upset by this and will not want to be on the site.

The next best interview place is a private and neutral location that will not intimidate the witness, such as a conference room. An administrator's office is not an advised interview location because it may put the witness on guard.

Witnesses must understand that the fact-finding interview is not a disciplinary interview, and so this should be communicated to them prior to asking any questions.

Incident Reporting

To improve safety and health, all incidents must be reported. Incidents go unreported for several reasons. One reason is a safety program that includes monetary incentives for safety performance. In this case, an accident (and its associated report) results in the employee or department losing some or all of the quarterly cash/bonus due to the incident. Because of this, an employee may seek to avoid monetary loss, ridicule, or embarrassment and instead choose to hide the incident by failing to report it. Such chains of events render safety programs ineffective and counterintuitive with respect to good safety culture behaviors and the goals of continual improvement.

Incident reporting and investigation are an after-the-fact response to an event or accident. The primary purpose of reporting and investigation is to determine the root cause and prevent the incident from recurring. Part of awareness in reporting incidents is knowing what should be reported. Aside from events where injury and loss have occurred, close calls and near-miss incidents must be reported as well. Such situations (close calls) serve as indicators that an accident will eventually occur. Effective site safety plans should outline incident investigation procedures and what they entail. Any corrective actions implemented after the incident should be tracked, monitored, and measured to determine the level of effectiveness in preventing further incidents. Such actions are critical to continual improvement efforts. The four general steps for incident investigations are as follows:

Describe the Scene

- Guard the scene via equipment (e.g. cones, barriers) to prevent important materials from being disturbed or removed.
- Record facts such as the time, date, and location of the incident; name(s) of the injured; a description of the severity of the injury; and the names of witnesses.
- Take photos and/or videos of evidence at the scene and information that could be critical for analysis.

Data Collection

- Take statements and conduct interviews as soon as possible (utilizing recent memory to acquire more accurate information). Practice greater caution when interviewing the injured person(s).
- Review equipment and maintenance inspection programs, work processes and procedures, and operating manuals.
- Review records for similar incidents and follow-up actions.
- Review processes relevant to worker training, discipline, and enforcement records.

Determine the Root Cause

- Apply root cause analysis through a series of "*Why?*" questions methodology; proceeding with *if not*, or *if so,* then ... ("Why?" repeating) to determine the origin or cause.
- Apply root cause analysis to detect failures in procedures and apply actions to program and process deficiencies, not to assign personal fault.

Apply Corrective Actions

- Present adequate causes for corrective actions to upper management.
- Avoid presenting less significant causal factors to management (e.g., not using PPE).

- Target and modify program-level elements (e.g., equipment selection, procedures, maintenance programs) for corrective actions and implementations.
- Follow up on corrective actions with tracking and monitoring to determine effectiveness in preventing further incidents.

The incident report will consist of a summary and narrative. The narrative includes findings, conclusions, and recommendations for corrective action. The summary is the most important component of the report because it identifies the need for the report; the summary must be written in a manner that will encourage the reviewer to read the narrative. An ASP will encounter administrators who may not be in the habit of reading reports; an effective summary will increase the likelihood that the rest of the report will be reviewed.

Chain of Custody

There is a general duty to preserve evidence whenever it can be reasonably foreseen that the information may be relevant to the investigation. Formal chain-of-custody procedures will help to ensure that evidence is handled appropriately and hold employees responsible for doing so.

The starting point for chain-of-custody is when the evidence is documented for context, time, conditions, and anything else material to the particular evidence. For example, identifying photo orientation and lighting would be useful for photo evidence.

Relevant and correct documentation and notation are of the utmost importance because evidence may be reviewed by parties that have never visited the site of the incident, and the evidence and witness interview transcripts may be the only sources of information for them. CAPA records are also critical, historical documents that should be readily available for reference.

All employees assuming custody should be required to sign a chain-of-custody document and should receive a receipt of the transfer. The number of employees handling the evidence should be kept to a minimum to reduce the likelihood of errors and mishaps in the transfer process.

Conducting Inspections and Audits

Inspections and Audits

Identifying and controlling hazards is a continuous process. When hazards are identified, they should be promptly and correctly eliminated. *Competent person* is a frequent term used in the construction industry, for example, and OSHA calls for at least one competent person to be present on each construction site. Defined by OSHA, a competent person is "… one who is capable of identifying existing and predictable hazards in the surroundings or working conditions which are unsanitary, hazardous, or dangerous to employees, and who has authorization to take prompt corrective measures to eliminate them." OSHA's definition applies mostly to foremen, but any sufficiently capable employee can be deemed a competent person and authorized to identify and correct hazards. So long as hazard control is taking place, any number of competent persons may conduct inspections, whether they are a general laborer, foreman, or safety professional.

OSHA's guidance regarding inspections is to conduct "frequent and regular" inspections, allowing the contractor to determine the frequency and regularity of inspections on the job site. If hazards are corrected in a timely and reliable manner, the contractor's inspection regimen is working as intended. If hazards still exist after an inspection is conducted, the contractor's follow-up regimen has failed.

OSHA does not require inspections to be documented but mandates that identified hazards must be corrected. The absence of hazards proves that inspections and successful hazard-control practices are taking place.

Documenting inspection and hazard-control activities will provide valuable information. Even when only very minor hazards are identified that do not require work to stop in order to correct and may not even be mentioned at a safety meeting, documenting observations will provide safety administrators with data that they will use to analyze trends and allocate resources more effectively in the future.

The Board of Certified Safety Professionals believes that hazard control is most effective when the immediate supervisor is responsible for correcting identified hazards. The immediate supervisor is most likely to know the particular work area, its hazards, and the employees in that work area. Sometimes the supervisor is an ASP, but usually the ASP will work in an advisory role to assist managers to maintain maximum safety at all times. While an ASP may be capable of correcting the hazard, there can be unintended consequences. Employees may observe the ASP, a third party to their operation, as correcting hazards in their workplace and assume that this is just a corporate formality. They may surmise that if the hazard is not important enough for their supervisor to deal with directly, then why should they pay it any thought? Therefore, it is advisable that all project management be intimately involved with the identification, prioritization, and correction of all hazards.

Equipment Inspection

Scheduled inspections and preventive maintenance procedures detect and address equipment deficiencies and potential hazards before an incident occurs. This provides a higher level of safety to employees and positively impacts quality, production, and profit. Inspections are performed by personnel such as managers, supervisors, and maintenance workers and should be led and monitored by the safety professional. Inspection plans/programs are most effective when a combination of daily and scheduled inspection processes is used and properly enforced.

Organizations should emphasize the importance of routinely inspecting machines and equipment. The purpose of these inspections is to identify potential hazards and correct them through preventive maintenance before an incident occurs. Beyond improving employee safety and providing benefits to quality and production, inspections can positively impact performance, efficiency, and profitability. They also can be performed by different members of an organization, including managers, supervisors, maintenance, and medical personnel, and they can be led and continuously monitored by the safety professional. Inspections can be done on a **continuous** (daily) basis or on a **scheduled** (interval) basis. Inspection programs are most effective when using a combination of both. Prior to implementing an inspection program, an organization should determine the following:

- What requires an inspection?
- Which parts should be examined?
- The appropriate physical condition of items for inspection
- The frequency of inspection
- Which personnel will conduct the inspection?

Evaluating Cost, Schedule, Performance, and Project Risk

Evaluation of Cost, Schedule, Performance, and Risk

Project management involves the planning and organization of a project and its resources. A project plan will outline the approach a team will take and helps the stakeholders/team document decisions in regard to the scope, objective, schedule, and resources/deliverables. There are several components of project management, including the creation of project goals, the formulation of project teams, and the setting of core tasks/priorities to ensure that a team completes a project. Projects consist of multiple phases. A life cycle sets the beginning and end of a project. Scope, time, and cost management make up the core of project management. Other main areas of project management include integration, quality, communication, risk, procurement, human resources, and evaluation of new services, products, and technology.

A primary task for a project manager is to evaluate costs. **Cost management** requires estimating the costs of a project and uses technology to measure productivity and costs. Time, money, and expertise must be considered when evaluating the costs. Cost management is vital for every stage of the life cycle. A **budget matrix** is typically developed to keep track of project costs. Estimate templates can be used to estimate the cost at all task levels. These templates help calculate the cost associated with individual costs, such as the labor, task length, the hourly rate of the worker, and needed materials. At the final stage of a budget cycle, there is a budget evaluation, or a final assessment of whether resources will be used effectively for a given task. The budget evaluation will outline different options and designs for a particular task and will help identify repeatable costs. The total cost of the evaluation is the sum of the individual expenses.

Scheduling measures the progress of a project, outlines how a team member contributes to a project, and shows dependencies. A task or activity that cannot be completed until another task is completed is called a dependency. Schedules define what task needs to be completed and determine which team member or point of contact is responsible for the work. Schedules will show the life cycle of a project and break down specific deliverables and key points.

Performance management is a continual communication process between the employee and supervisor that occurs throughout the year. The employee and supervisor of the organization will work together to identify workplace goals, set objectives, clarify expectations, review performance results, and give each other feedback.

Risk management encompasses the identification, assessment, and control of threats in regard to an organization's capital/earnings. The source of the threats can be attributed to errors in strategic management, natural disasters, accidents, legal liabilities, and financial uncertainty. Risk management strategies are also important to companies that use digital technology, which can undergo data-related and IT security threats. The organization will use risk management to prepare for the unexpected by reducing risks and extra costs.

Ergonomics

Fitness for Duty

Ergonomics, or the study and design of the work environment to address physical demands placed on employees, is yet another area addressed by OSHA. In the workplace, ergonomics deals with elements such as lighting, placement of controls, equipment layout, and fatigue. OSHA examines work-related injuries that result from repetitive stress and repetitive motion, such as carpal tunnel syndrome. These are also known as cumulative trauma disorders. These workplace injuries may be reduced by redesigning workstations and improving workplace environments.

Construction activities present many ergonomic hazards that can result in crippling injuries. These hazards stem from: using hand and power tools; processes that require static postures and positioning; improper body mechanics while pushing, pulling, or lifting; repetitive stress; and elements between workers and material that cause slippage as a result of friction and grip loss.

Ergonomics refers to the ability of a person to fully utilize a product while maintaining maximum safety, efficiency, and comfort. Ergonomic risk factors in the workplace can lead to musculoskeletal disorders such as carpal tunnel syndrome, rotator cuff injuries, muscle strains, and lower back injuries. In order to reduce these risks, employers should evaluate workplace ergonomics and educate employees about potential issues. An ergonomic evaluation tests a product to determine its ease of use and potential safety risks. When employers identify and address ergonomic concerns in the workplace, they protect their workers and likely prevent serious injuries.

Both behavioral and psychological factors can greatly influence the ergonomics of the worker's daily flow and output. **Behavioral factors** can include a worker's physical ability to perform work tasks (e.g., the ability to lift heavy items or sit for long periods of time), a worker's overall physical health, and the way a single worker interacts with other people or equipment in their work environment. **Psychological influences** include a worker's mental health status; their attitudes and perceptions of their own abilities to perform required job tasks; their general comfort level during the course of the workday; and their personal level of agreeableness, work ethic, or cooperation in a work environment. Finally, cognitive factors, such as the ability to learn new skills, interface with technology, or remember important details, can play a role in both behavioral and psychological influences.

Stressors

Repeated exposure to ergonomic stressors is a risk to worker health and safety. **Ergonomic stressors** are those that cause undue physical, mental, or emotional strain on workers.

Environmental

The field of ergonomics mainly focuses on how stressors in the physical working environment affect the well-being and safety of workers and how to reduce or eliminate these stressors. Often, implementation focuses on reducing the risk of musculoskeletal disorders and injury from repetitive physical stress; however, someone who experiences physical stress or workplace danger or who is compromised biomechanically at work is also likely to feel mental and emotional stress. Therefore, solutions may incorporate ways to resolve mental and emotional stressors as well. Ergonomic stressors vary by job role; for example, someone who primarily works all day at a desk will experience different stressors than someone who primarily performs manual labor.

Lights

Lighting as it relates to ergonomics consists of providing sufficient light for a person in a workplace without putting too much stress on the eyes. Too much or too little light may hinder performance and cause undue stress. It is best to provide sufficient lighting based on the environment. For example, if there are numerous large windows, then less intensive lighting may be permissible. However, the goal is to eliminate dark areas within in the workplace without creating an imbalance of light. Reducing glare is another way to reduce stress on the eyes and decrease the likelihood of fatigue. Glare reduction may be addressed by installing blinds, matching computer monitor brightness to room brightness, swapping in lower intensity light bulbs, and implementing other preventative measures.

Noise

Noise is undesired sound. Sound waves extend from a noise source. Depending on the duration or the severity of the noise, it may cause a person harm, such as hearing loss. However, harm may be prevented with OSHA's hearing conservation program. This standard may be found in 29 CFR 1910.95. Employees subject to this standard must submit to audiometric testing. The test is conducted by a certified specialist within 6 months of the initial noise/environment interaction measuring more than 85 dBA. Before an audiogram baseline may be established, an employee must not be subject to excessive noise for 14 hours prior to testing. An employee qualifies for a hearing conservation program if he or she is subject to noise exposure of 85 or more dBA. Different types of instruments may measure exposure to sound. A type 1 sound-level meter shows ±1 dB accuracy for exact settings. However, the minimum standard for exposure determinations, a type 2 sound level-meter, shows ±2 dB accuracy for general purposes.

Other Conditions

Common ergonomic stressors include awkward positioning, such as needing to work in unique spaces during the workday (e.g., the crawlspace of a home); extreme environmental temperatures (e.g., outdoor daytime work at the peak of summer); or poor fit between necessary equipment and the worker using the equipment. Repetitive stressors are the most common cause of ergonomic hazards that result in injury and can be found in most jobs. **Repetitive stress injuries** occur when a worker performs similar tasks day after day. The regular, cumulative physical demands of a job on the worker's body are likely to have a negative impact over time. Therefore, jobs should have mitigation tactics that are tailored not only to the tasks of the job but also to the individual performing the job. For example, two employees may both work on a production line where they package materials and shelve the sealed boxes. While all employees on this production line may benefit from learning proper lifting techniques to shelve the sealed boxes, each individual employee may have different ergonomic needs based on unique physical factors such as age, height, strength, personal physical limitations, arm reach, and so on.

Risk Factors

Some occupations are at a higher risk for ergonomic hazards due to a highly physical nature of work. These occupations include most clinical healthcare and emergency service positions, manufacturing positions, mechanical workers, commercial and public vehicle drivers, and manual laborers. Risk factors can be reduced by identifying workers who are especially vulnerable and providing additional support (e.g., through extended rest periods); cross-training workers and rotating responsibilities to avoid repetition; providing tools to reduce known ergonomic stressors; providing adequate rest for workers; creating a culture of open communication about potential safety hazards; and providing physical fitness, well-being, and work technique training to cultivate a strong, resilient workforce.

Repetition

Risk factors for ergonomic hazards typically result from the way work is performed, although they may also result from the environment in which work occurs. A common risk factor for ergonomic stress is repetitive job responsibilities, in which the worker repeatedly and regularly uses the same motions and muscle groups. Repetitive actions are found in virtually any kind of job.

Force

Another risk factor is the regular use of human force in a job, such as in duties that require heavy lifting, carrying items, gripping tools or equipment, or pushing or pulling materials. This places strain on large portions of the musculoskeletal system. When a job requires forceful movements, workers are at a higher risk of injury if they are not rested, have low fitness, or have low muscular strength.

Posture

The worker's posture on the job can also be a critical risk factor. Workers who have to hold awkward positions that place unnatural loads on the spine, hips, or other supportive skeletal structures may develop muscle sprains or strains over time. In the short term they may experience unpleasant side effects, such as aches, pains, fatigue, and irritability. The use of tools or equipment can help mitigate postural risk factors. For example, many desk workers can benefit from headsets (rather than spending long periods of time holding a phone) and stand-up desks, while mechanical workers can benefit from tools that limit awkward positioning around machinery.

Vibration

Occupational vibration can be hazardous due to the repetitive and forceful shaking that can occur across the body or upon a limb. This most commonly occurs when using a vibrating tool, such as a jack hammer or chainsaw. Heavy vibration to the hand and arm is most likely to result in ergonomic injury.

Many power and pneumatic tools can produce hazardous vibrations. These tools include air and electric hammers, hammer drills, tampers, chisels, chain saws, and grinders. Long-term use and exposure can result in **Hand-Arm Vibration Syndrome (HAVS)**. Symptoms of HAVS include circulation disruption, hand and finger numbness with loss of dexterity, muscle and bone disturbances, and compromised performance in detailed tasks. NIOSH has threshold limits for hazardous vibration. Limit parameters are based on time and acceleration (duration of exposure and directional frequency of vibration). A control measure for hazardous vibration is to provide anti-vibration gloves as PPE. Factors that limit their effectiveness are glove type, hand size, and grip strength among different workers.

Work Design

For addressing hazards, the underlying **principle of ergonomics** is to fit the job needs to the abilities of the worker. OSHA currently has no ergonomic standards regarding construction and CFR 1926 but, as with many situations, concerns can be cited through *OSHA's* **Section 5(a) (1) General Duty Clause** which states, "employers are required to provide their employees with a place of employment that is free from recognizable hazards that are causing or likely to cause death or serious harm to employees."

Ergonomic hazards can be addressed through the development of an effective program. Such a program should be geared toward ergonomics and contain the following core elements:

- Management commitment with clear objectives and performance reviews
- Employee involvement with communication and feedback
- Detection of hazards through JSAs
- Establishment of hierarchical controls
- Medical management related to ergonomic issues
- Training, monitoring, and program enforcement

Conducting JSAs is vital for applying ergonomic principles to construction processes. The presence and variety of changing elements require that job analyses be performed in conjunction with other procedures that acquire more information than standard JSAs. Elements should include the following:

- Observing the worker performing the task
- Taking photos and videotaping
- Conducting interviews with workers
- Conducting surveys and questionnaires
- Making calculations of force requirements, reach distance, and material positioning

Once the initial hazard survey is complete, the ASP will have the information necessary to determine the most effective measures including equipment adaptations, alternate material handling practices, and their worksite modifications. It is necessary for the ASP to be present in the worksite in order to mentor workers and ensure the agreed-to ergonomic procedures are adhered to. All worksite applications and procedural changes must be documented, disseminated, and properly communicated to all levels of personnel with all available methods. When an ergonomic problem or hazard is encountered on the worksite, its identification is followed by assessment and mitigation. Solutions can be implemented according to the priority of the issue. Immediate hazards that do not constitute major process changes should be made immediately. More difficult hazards and problems that involve process changes should be associated with a CAPA plan.

When an ergonomic problem or hazard is encountered on the worksite, its identification is followed by assessment and mitigation. Assessment provides the information needed to prioritize the hazard and allocate the resources deemed relative to its risk factor (on a standard risk matrix, this will be low, medium, or high). Mitigation of this hazard will assign the control appropriate (elimination, substitution, engineering, administrative, or PPE) to close the hazard out. Again, whether or not a formal CAPA plan is required will need to be decided by all project management personnel and the ASP.

Material Handling

Material handling incidents constitute almost a third of construction workers' compensation claims and a full quarter of all industry claims. The sobering number and frequency of incidents speak to a common denominator between the similarity of these injuries and the repetitive nature of these occurrences. In short, most of these are known and can be mitigated, if not avoided altogether.

Work-related musculoskeletal disorders (WMSD) typically result from the repetitive and task-specific movements associated with construction and other heavy industries. Where injuries like sprains and minor lacerations would have been accepted in the past as an inherent part of the workplace, effort and attention are being directed to reduce the level of worksite incidents to **As Low as Reasonably Practicable (ALARP)** levels.

Manual

Alternative task methods and tool modifications have helped to alleviate the manual handling of heavier materials and provide improved ergonomics that prevent bodily positions typically associated with situational or chronic injuries. The extended rebar tie tool, the kneeling creeper, and the extended screw gun are examples of more recent tool modifications designed to alleviate potential hazards.

Manual material handling on construction sites continues to be a challenge. In some circumstances, limited working areas created by traffic zones, equipment obstruction, or confined spaces constrict worker movement and maneuvering space; this in turn hinders the capacity to use a neutral body position or mechanical aids for pushing, lifting, pulling, or carrying materials.

Lifting Devices

Although many mechanical aids for labor have been devised and employed at the worksite, lifting is still a mainly manual activity that is frequently performed. Poor lifting habits are a workplace staple. Workers will naturally lift an object in a manner that is most convenient, which precludes the proper body placement for optimum lifting (humans have spines, and cranes have booms; load capacity is reduced, and strain increases in both as they extend horizontally [boom out], torque, and deviate from a straight-up position). Proper lifting will prevent spinal column injury by avoiding lifts that are away from the body and that involve a twist while lifting. Children naturally lift properly. Their lack of musculoskeletal development forces them to squat down and position the load close to the spinal column before they lift. This, as it turns out, happens to be the proper lifting technique. Standard recommendations regarding proper lifting include the following:

- When lifting a load, it should be kept close to the body (within ten inches).
- Do not twist or torque the body during lifting an object or setting it down.
- Keep the back straight, and lift with the legs.
- Use a solid, two-hand grip.

Injuries resulting from forces on the neck and spine can be greatly reduced by mechanizing the material handling process. Appropriated forklifts and carts can transport material from trailers or stockpiles to areas where the items are required. Some are even equipped to lift materials to required heights. Such equipment can reduce slip, trip, and fall injuries by reducing or eliminating carrying distances and traversing over uneven or slippery ground surfaces. Conveyor systems also can be added to the process, reducing the workload associated with manual tasks and minimizing risk.

Powered Equipment

With or without a standard hand truck (dolly), lifting and carrying can place heavy loads on the shoulders and spine. Powered dollies are an improvement over traditional dollies with the removal of awkward postures and force requirements. This equipment also is available with stair climbing features and adjustable handles.

Delivery trucks can be equipped with hydraulic lift gates, which are designed for loading and unloading materials and equipment. This eliminates the hazards associated with workers climbing on and off the rear of the truck, as well as the handling of other loading equipment.

Adjustable scaffolding provides continuous height adjustment. This allows brick and block layers to work from waist level, which minimizes back pain stemming from postures associated with bending and repetition. This system also enables planking to stay in place, eliminating the need to change it at different working levels.

Work Practice Controls

Work practice controls are methods that reduce ergonomic hazards and minimize risks to the workers. Continuous exposure to ergonomic hazards is likely to result in musculoskeletal disorders or burnout. Work practice controls identify and reduce, or eliminate, processes that can ultimately risk worker well-being and safety. Ergonomic controls primarily intend to eliminate frequency or repetition of a task, excessive force or strain by the worker, and unnatural postural movements. However, sometimes these risks are inherent to the job and unavoidable in the work process. For example, many data-oriented jobs require extensive periods of time typing on computers, looking at computer monitors, and clicking on a computer mouse in order to produce required process outcomes.

Job Rotation

Job rotation is a form of control in which workers rotate the tasks they are completing so that no single worker is doing the same task for the duration of a work shift. For example, in order to reduce time spent typing, looking at a computer screen, or clicking on a mouse, a group of data analysts may rotate the duties of data entry, printing and assembling copies, and holding face-to-face meetings to present data findings. Regular breaks in which workers stretch, rest, move their limbs across various planes of motion, or otherwise stop a repetitive work activity can also control ergonomic risks. Additional proactive controls include training and education on proper body mechanics (e.g., lifting techniques for heavy items); education concerning personal health and care, especially in physically demanding jobs; and adequate time off when needed by the worker.

Work Hardening

Work hardening is a control that occurs in response to a workplace musculoskeletal injury. When a worker is injured, they may have already been susceptible to the injury (e.g., sedentary lifestyle for a physically active job); even if they were not vulnerable to injury, they may lose fitness and mobility as they recover from the injury. Work hardening is a customized form of physical therapy that conditions the worker for their specific job role. Programs focus on increasing mobility, endurance, and strength as it relates to the job task. For example, a worker who is required to lift heavy objects may use a program that focuses on strengthening the lower body and torso while incorporating healthy lifting techniques. A worker who sits at a desk most of the day may strengthen the upper back and arms and stretch the chest muscles while also learning ergonomic sitting techniques to reduce spinal discomfort. The following graphic shows the impact of different controls on ergonomics.

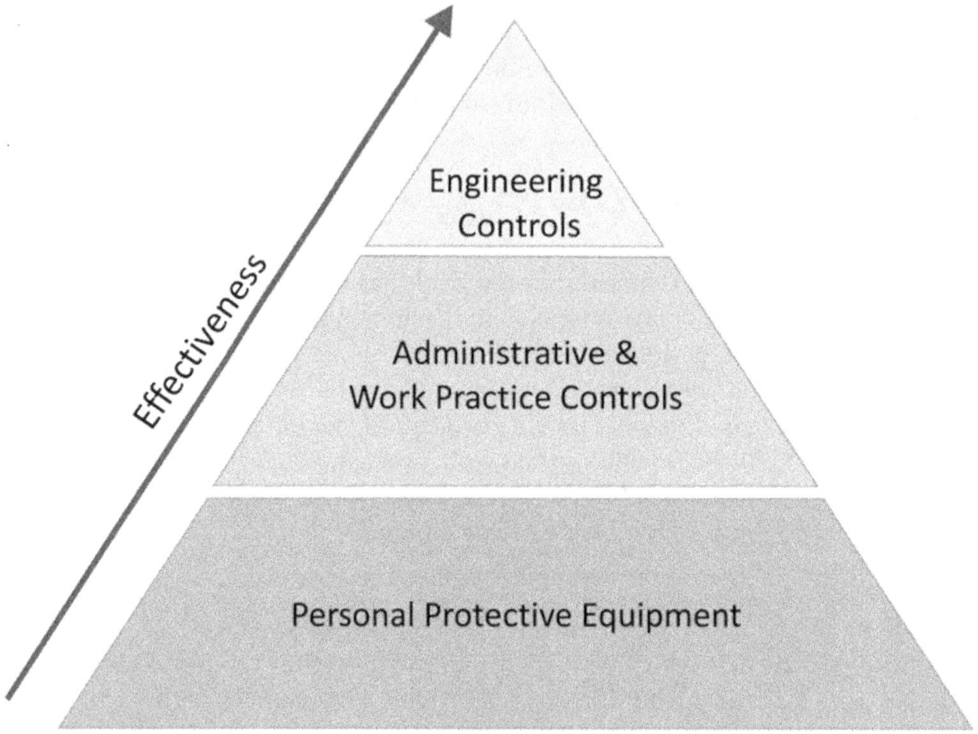

Early Symptom Intervention

Early symptom intervention allows for companies to discover early signs of potential injuries and prevent the injuries from worsening. Generally, early signs of such injuries include the worker experiencing fatigue and discomfort. Companies should encourage personnel to report these signs. The purpose of the intervention is to prevent harm, not to treat it. This prevents an employee from experiencing painful musculoskeletal disorders and prevents the company from being subject to litigation. Some preventatives that companies have implemented include stretch times, proper lifting techniques, proper posture and form for work activities, and other actions that decrease the risk of injury.

Using Qualitative and Quantitative Analysis Methods

Quantitative Methods

Quantitative ergonomic assessment methods provide useful, objective data about health and safety risks in a specific workplace. Quantitative methods focus on numerical metrics that provide a picture of health and safety information. Common quantitative data sources in ergonomics include reports such as health insurance claims for musculoskeletal complaints, workers' compensation claims, and total number of falls. Some methods provide a baseline assessment that allow an organization to determine where its health and safety standards currently are in order to identify strengths as well as opportunities for improvement. Other quantitative methods measure a specific metric that has been identified as an opportunity for improvement. Tests of change may be implemented, and then the metric is monitored in real time to see whether a specific risk threat is increasing or decreasing.

Anthropometry

Anthropometry is a measurement system that analyzes a person's various physical attributes (e.g., hand size, arm length) and how these attributes interact with various components of a workplace. Anthropometry provides data about a specific workforce that helps to design that workforce's environment, equipment, tools, uniforms, protective gear, and technological devices. For example, the size of a person's hand and arm will determine parameters for power tool usage, as tool vibration can pose an ergonomic hazard. Anthropometry is an important part of ergonomics as workforces become more diverse in terms of gender, age, physical stature, and so on.

NIOSH Lift Equation

The National Institute for Occupational Safety and Health (NIOSH) is a leading public research agency in the United States that focuses on worker well-being and safety. The **NIOSH Lift Equation** is considered a reliable and valid measurement tool to assess risks for humans who are lifting objects. The Lift Equation analyzes several spatial variables to provide a Recommended Weight Limit for a specific lifting event. Objects that are heavier than the Recommended Weight Limit are likely to increase musculoskeletal risk for the average healthy worker.

Variables that are considered in the Lift Equation include where the item that is to be lifted is in location, both horizontally and vertically, to the person who is lifting; the distance the object is being lifted; whether any twisting motions are required; how often the item must be lifted in a single event; and what level of grip is required on the object. The NIOSH Lift Equation also assesses risk by providing a Lifting Index associated with the task. The Lifting Index is determined by dividing the Recommended Weight Limit by the weight of the object that is being lifted to produce a value that is higher or lower than 1.0. Tasks that score higher than 1.0 are considered high-risk; these tasks should be further evaluated and possibly re-designed to lower the Lifting Index score.

Qualitative Methods

Qualitative ergonomic assessment methods provide subjective yet informative data about health and safety risks in a workplace. Often, this information comes from a safety professional's informed review, conversations with workers in the environment that is being assessed, or anonymous organization-wide surveys. Qualitative data collection regarding workplace ergonomics may focus on the worker's perspective and insight about working conditions; personal reports of pain, fatigue, and discomfort while working; and personal safety concerns.

The **Rapid Upper Limb Assessment (RULA)** was developed by ergonomists from the University of Nottingham, a public university in England. The RULA aims to be a simple qualitative assessment to determine the risk of musculoskeletal injury to specific body regions (e.g., wrist, shoulder) when performing a specific task. An external evaluator interviews the worker who is being assessed to gather information about the worker's job and various tasks. Based on worker input and the evaluator's assessment, the most strenuous, lengthy, forceful, and frequent job tasks are assessed for risk. Tasks are scored based on several factors, including weight of any items that are held; how a body part is positioned in relation to the midline, torso, or a major skeletal joint; and postural impacts. Scores are then added up; tasks that score 4 or lower are considered low risk, while tasks that score 5 or higher are considered high-risk and likely need to be redesigned. Because an external evaluator assigns the risk score based on their personal knowledge, opinion, and observation, this is considered a qualitative assessment.

The **Rapid Entire Body Assessment (REBA)**, also from the University of Nottingham, assesses musculoskeletal postural risk to the upper and lower parts of the body during specific tasks. Movement and risk are assessed across different planes of motion, across changing postures, across quick movements, and in situations where gripping an object affects posture. The REBA assesses the left and right sides of the body individually, and both sides might not be assessed during a single evaluation. Like the RULA, an external evaluator interviews a worker about their movements during their job tasks. Based on worker input and evaluator observation, postures that are considered the most strenuous, experiencing the most force, or held for extended periods of time are assessed. A risk is scored on several variables, such as flexion and extension of the spine in various positions, load upon joints, sudden exertion, rapid movements, and grip positioning. Scores are added and fall within a range of 1 to 15, with 1 being low risk and 15 being the highest risk.

Fire Prevention and Protection

Chemical Hazards

Chemical fires can occur when structural changes in a chemical compound arise due to force or pressure; as the compound experiences these structural changes, excess heat energy is released into the surrounding air. Chemical fires are likely to cause explosions, which increase the dangers to human life. In addition to burns, people who are near a chemical fire may find themselves victim to shockwaves, inhalation of noxious chemicals, and trauma from projectiles. An initial chemical explosion is almost always likely to cause secondary explosions, which are generally more dangerous than the initial explosion due to additional flammable particulates that are released and dispersed from the primary explosion.

Industrial workplaces that utilize chemical combustion processes to create different chemicals, manufactured products, or energy sources are at a higher risk of experiencing chemical fires. In these workplaces, areas of a site can become contaminated with various dusts and particles that can increase pressure in an area, react with one another, and inadvertently become explosive or flammable. High-risk industries include energy production plants, chemical plants, pharmaceutical plants, pesticide and fertilizer plants, plastics plants, and most industrial agriculture sites. These industries use flammable materials, such as gasoline, ether, petroleum, glycerol, and chloric acids. These compounds are flammable on their own but can become more dangerous when they are combined with other common production materials.

Finally, chemical fires are not limited to industrial production plants. Chemical fires can occur in residential settings during routine activities like cooking. For example, household equipment items such as lawnmowers require gasoline to function. Poor storage mechanisms can cause flammable vapors to escape and interact with sources of heat that can result in a fire.

Auto Ignition and Flash Point

Flammable materials are combustible materials that will burn in air at environmental temperatures. The lowest temperature that will cause a chemical to ignite, without an available ignition source, is called an auto-ignition temperature. The lowest temperature at which vapor, over a material, can ignite in the presence of an ignition source is called the **flash point**. Fire is a combustion process that involves the oxidation of a material in oxygen. The chemical reaction of oxygen and fuel releases heat. Fuel is a substance that can react with another substance, such that it releases heat energy. Wood, paper, and chemicals are fuels. A chemical reaction occurs in the presence of heat, oxygen, and fuel, as indicated by the fire triangle, or fire tetrahedron model.

The fire can be extinguished if oxygen, heat, or fuel is removed. Cooling down the fire with water or a fire retardant can extinguish it. Due to the nature of the fuel, which can consist of various metals or chemicals, firefighters must use different types of fire extinguishers.

Electrical Hazards

Electrical issues are a common source of fires in the United States due to the wide-ranging use of electricity in commercial and residential structures. Electric currents form due to the flow of electrons, which hold energy. When electrical currents are impeded in some way, the kinetic energy in the current begins to increase and cause heat. While most modern circuitry includes mitigation and control methods to disperse energy and heat should an impedance occur, accidents can still happen.

OSHA standards require and assume that only qualified employees perform electrical work. A similar position is taken by the practices and procedures outlined in **NFPA 70E** (not an OSHA standard). Top electrical hazards include work near overhead power lines, electrical equipment and **lockout/tagout (LOTO)** issues (verifying machines are properly turned off at the end of each shift), wiring, and lighting fixtures. Providing measures of protection for electrical hazards are the responsibility of the employer. This begins with providing adequate training for employees that should consist of: evaluating hazardous job conditions, work processes, and precautionary items; energy isolation controls; approach distances for low-voltage (50-600 V) and high-voltage (> 600 V) panels; and appropriate PPE.

Control items include using double-insulated power tools, rubber coated non-conductive hand tools, and materials used near energized power lines and equipment. Gloves, boots, clothing and coverings, electrically-rated hard-hats, and other protective equipment must be inspected and meet their associated **ANSI** requirements. PPE selection is dependent upon several factors including voltage, current, distance, and duration of exposure. Testing meters must be approved and of an adequate voltage rating. A vital measure of protection includes the installation of **Ground Fault Circuit Interrupter (GFCI)** protection in wet areas and with all tools and equipment deemed temporary. Inspections of tools and equipment should be performed to detect hazards such as torn and damaged insulation, exposed wiring and junctions, shortened or altered cords, and circuit overload.

Static Electricity

Static electricity results from friction between two objects; as two objects rub against each other, electron balance between the two objects is not necessarily maintained. The object that has weaker electron bonds (normally due to the material it is made out of) will release more electrons than the object with stronger electron bonds. Therefore, it will end up with fewer total electrons and hold a positive charge, while the other object will have more electrons and a negative charge. The object with more electrons will hold the charge until it comes into contact with something that is a good conductor of electricity (e.g., metal) and then cause a sudden discharge of electrons. This can result in a noticeable spark. If this occurs near a flammable item, a fire or explosion can easily occur.

Surge

Electric surges occur when current flow suddenly speeds up along a circuit. This can happen from lightning strikes, plugging in too many appliances into too few outlets, or downed trees that impede balanced electrical flows into buildings. The circuit may not be able to handle the surge, especially if mitigating controls are worn or faulty. Electrical surges can concentrate electricity in an area and result in wires, appliances, or other objects heating up and catching on fire. Electrical fires and explosions most commonly occur as a result of old or improperly maintained circuits, improper use of electrical machinery, or user error.

Arc Flash

Flashover, or **arc flash**, produces heat and light due to the presence of an arc fault, an electrical discharge resulting from an air-to-ground connection. When the tungsten filament within an incandescent light bulb breaks or burns out, an arc is temporarily sustained between the broken filaments. The bluish flash that is often observed is due to the formation of plasma. Built-in fuses are typically equipped with house light bulbs to prevent an arc flash. Electrical devices exceeding 400 volts have enough capacity to create an arc flash, which can ignite a worker's clothing and cause arc burns. Arc burns don't require direct contact with electrical sources and are the result of high temperatures created by electrical arcs near the body.

Ground-Fault Current Interruption (GFCI)

OSHA's 1910 General Industry Standards require that GFCI be provided for all installations in bathrooms and on rooftops, but in construction activity, GFCI is required everywhere with one exception. The only alternative to GFCI in construction activity is by using a formal "assured equipment grounding conductor program." However, this is not a common practice and should not be considered a viable alternative to GFCI in most cases. An assured equipment grounding conductor program requires a substantial amount of documentation and equipment testing that, in most cases, will cost much more than the potential savings of foregoing the comprehensive use of GFCI-protected devices on a construction site.

A GFCI-protected electrical device will interrupt current when the device detects a disparity between the hot and neutral return wires. The amount of current in the wiring should always be the same in both directions; when it is not, a significant electrical hazard exists.

The difference in current between the two wires is not lost. Instead, the excess current goes to ground. When functioning properly, a GFCI device will shut off current when the difference in the two wires is between 5 and 7 milliamps. The interruption will take place within 1/40 of a second (0.025 second). GFCI devices should be inspected and tested frequently to ensure they are functioning as intended; GFCI testers can be readily acquired in hardware stores and large retailers.

A common misconception is that GFCI does not provide protection for non-grounded double-insulated tools because they do not have a ground prong. This is incorrect because the ground wire plays no part in the function of the GFCI circuitry; rather, the hot and neutral return wires are affected, so double-insulted tools will work with GFCI devices.

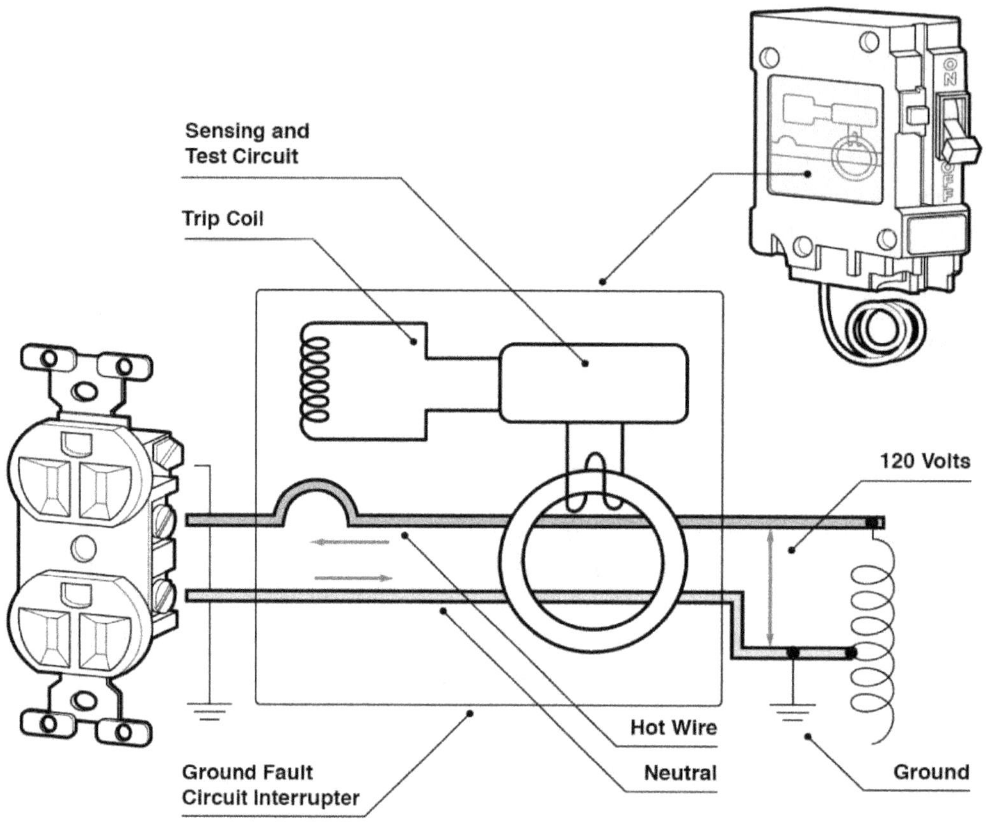

A GFCI-protected electrical device will interrupt current when the device detects a disparity between the hot and neutral return wires.

The type of GFCI installation is up to the contractor, who can choose to have every single receptacle equipped with GFCI, use a converter box, or employ a centralized circuitry system that will provide GFCI protection to all electrical tools plugged into the system. As with all electrical devices, the system employed must be inspected and tested to ensure safe operation and should be removed from service when deficiencies are identified.

Grounding and Bonding

The reason to have a continuous path to ground is to reduce the likelihood of current passing through a body if it somehow escapes the protection afforded by insulation. The ground prong on a power cord helps the current to get to ground as efficiently as possible and without employees being seriously harmed in the process. For this reason, a ground prong should never be removed from an electrical cord. Removing the ground prong will eliminate a component of the power cord intended to make it more likely that a worker will survive an electrical incident. However, this is a fairly common practice. OSHA frequently cites this violation of its electrical safety standards.

Grounding tools and equipment during construction is mandatory with one exception: double-insulated tools. Instead of a ground prong (third wire) on the power cord, a double-insulated tool will have an extra layer of insulation since the insulation alone is the only protection available. Double-insulated tools are indicated by a symbol: a square inside a square. They must have this symbol on them; it is the only way to readily identify a tool with no ground prong as being double insulated.

Older tools that are neither grounded nor double insulated that are encountered on construction sites should be removed from service immediately. OSHA allows no exemptions for older power tools or equipment manufactured prior to the enactment of the 1926 standards in 1971; they cannot be "grandfathered" in.

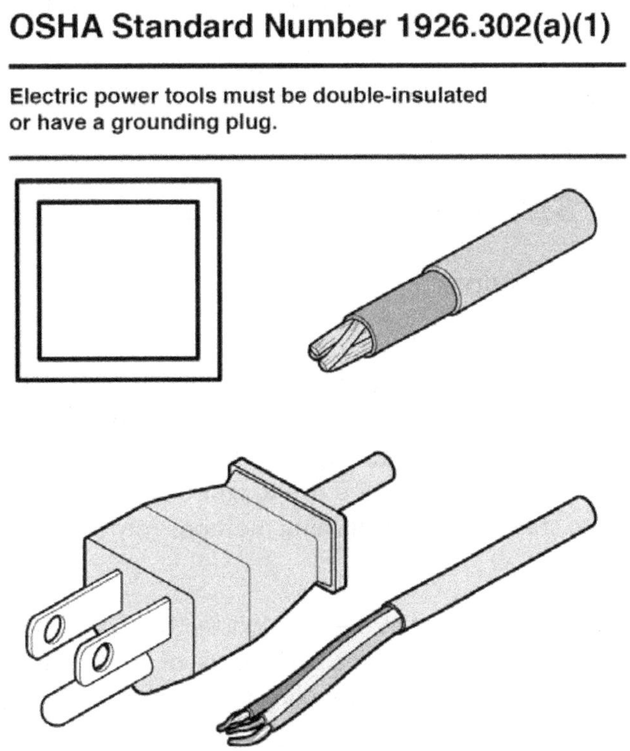

Hot Work

Welding, Cutting, and Brazing

Hot work is work that produces heat, fire, or sparks; examples include burning, cutting, welding, brazing, grinding, and soldering. This type of work presents many hazards from burning or electrical shock.

Hot work produces heat, sparks, or flame. Hot work presents the risk of burns from contact with hot tools or molten metal. Sparks can fly up to 35 feet and get caught on clothing, work surfaces, and unprotected openings in floors, walls, or construction material. Workers are at risk of being burned if flying sparks ignite when they come into contact with flammable materials on the work site such as wood, paper, soiled rags, clothing, dust, flammable liquid, or chemicals.

Hot work also presents the risk of workers being injured by explosions if flying sparks or hot tools come into contact with flammable gases that have leaked into the work space.

Heat-producing tools that run on electricity present the risk of electrical shock if the tool malfunctions.

To prevent serious injuries, workers should perform hot work only in a safe, well-ventilated location from which all combustibles have been removed. The work space should be enclosed in fire-resistant material, such as a welding curtain, to contain the heat and sparks. If workers cannot completely enclose the area, all combustibles should be protected by fire-resistant material such as welding blankets. To prevent explosions, the work space should be tested for flammable gas leaks before hot work is started.

To prevent electrical shock when using heat-producing tools, workers should make sure tools are properly grounded. Insulating mats should be used to prevent contact with fire and sparks.

Workers should always wear appropriate personal protective equipment for hot work. This includes welding gloves, eye and face protection, oil-free and fire-resistant clothing, and foot and leg protection such as tall leather boots with steel toes.

Fire Science and Combustible Dust

Fire Triangle and Fire Tetrahedon

Fire science is the field of study dedicated to understanding the prevention mechanisms, hazards, crisis responses, and emergency suppression techniques for fires. One basic model used in fire science is the **fire triangle**, which describes the variables needed to start a fire. These three variables are oxygen, heat, and a fuel source. The **fire tetrahedron** is a relatively recent expansion to the model that includes the exothermic chemical reactions that take place for a fire to ignite and continue burning. All four variables are needed in order for a fire to start, and extinguishing methods rely on removing one variable in order to stop the fire.

Extinguishing methods include reducing heat in the location, such as with cold water. However, whether water will extinguish a flame relies largely upon the fuel source of the fire. For example, grease fires are exacerbated by the addition of water due to the repelling forces of oil and water. Removing the fuel source is typically the most difficult part of extinguishing fires, and the other variables of the fire tetrahedron are typically addressed first. Fire blankets and hand-held fire extinguishers work to break down or remove oxygen in the fire. Some commercial fire extinguishers spray materials that stop the chemical reactions that result in fire.

Combustible Dust

The **combustible dust fire pentagon** describes the variables at play in a combustible dust explosion. Combustible dust explosions occur when combustible particulate matter is in an air space; they are especially dangerous because the particulates are often undetectable until an explosion occurs. In addition to oxygen, heat, and fuel, the combustible dust fire pentagon also accounts for dispersion and confinement. Dispersion refers to particulate suspension in the air, while confinement refers to how the air is enclosed. Confined spaces with combustible dust are more likely to result in an explosion because the reduced space allows for a significant increase in pressure. Fire and dust combustion prevention mechanisms focus on mitigating the interaction of the variables found in the fire tetrahedron and dust pentagon.

Fire Triangle

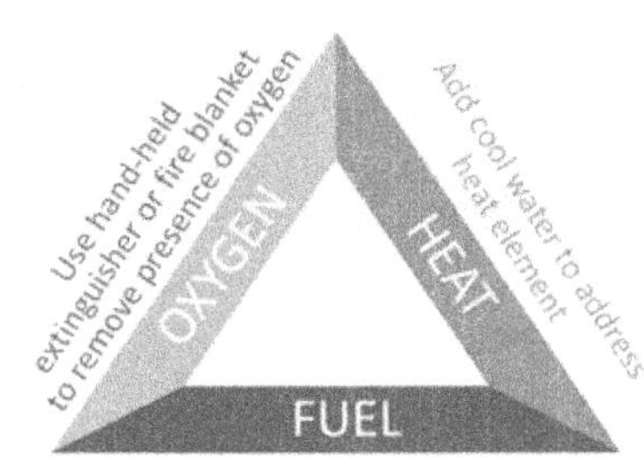

Fire Tetrahedron

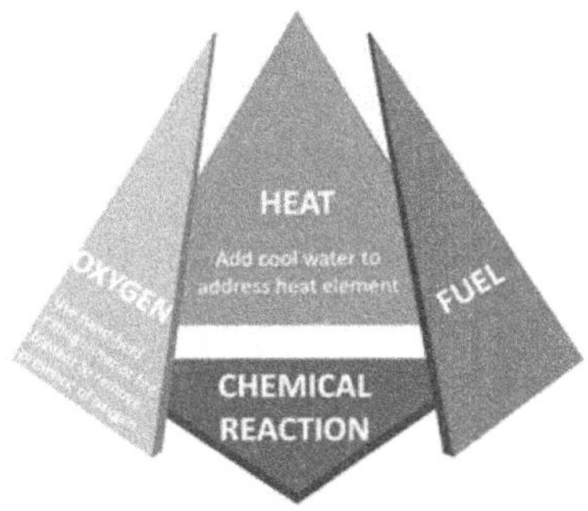

Dust Pentagon

Upper and Lower Explosive Limits

Testing and monitoring are performed for the presence of gases and other potentially hazardous or explosive atmospheric constituents. These are done to detect the presence of carbon monoxide, inherent or introduced toxins and gases, and oxygen deficiency and enrichment, which can impact hazardous atmospheres including the **lower explosive limit (LEL)** and **upper explosive limit (UEL)**. It is vital for the safety professional to know and consider factors of vapor density and the link between oxygen levels and potential modifying effects on LEL/UEL. Monitors are set with a safety factor so that an alarm is triggered if oxygen levels fall outside the range of 19.5 to 23.5% in atmosphere. They are also set at 10% and above for the LEL with respect to the presence of flammable gases, fuel-air mixing, combustion, and explosion.

Detection Systems

Fire detection mechanisms are an important way to control fires before they become catastrophic. In most commercial buildings, fire detection systems can be activated by any person who is nearby and notices hazardous conditions. These manual systems may then electronically sound alarms to alert occupants to evacuate the building, signal emergency response services, or set off sprinkler systems. Most autonomous detection mechanisms aim to sense excess heat or smoke rather than flames. Smoke is a precursor to fire, and the presence of actual flames makes a fire event considerably more dangerous and difficult to manage. Thermal detection systems detect both sudden increases in heat and abnormally high temperatures in a space.

Thermal detection systems are reliable and inexpensive; however, they typically do not work well in larger rooms where heat may be undetectable until flames are present. Thermal detection systems can be useful directly near heat sources (e.g., close to equipment that could pose a fire hazard). Smoke detectors are more commonly found in commercial and residential buildings; they work by detecting smoke. Well-functioning smoke detectors are highly advantageous over manual and thermal systems because human input is not needed; they can detect hazardous conditions for a fire before a flame forms. Finally, some fire detection systems operate by detecting actual flames. These are utilized primarily in industrial settings where heat and smoke are inherently found; therefore, other detection systems would result in continuous and potentially false alarms.

Suppression Systems, Fire Extinguishers, and Sprinkler Types

An ASP should be very familiar with basic fire safety principles, the classes of fires, and the types of commonly used extinguishing agents because many of the materials found in construction are flammable.

Fire extinguishers on construction sites should be inspected at least monthly. This is an NFPA (National Fire Protection Association) consensus standard that OSHA has adopted. The monthly inspection is a simple task that does not require an advanced knowledge of fire extinguishers, and many safety personnel will be sufficiently knowledgeable to be assigned this duty. During the monthly inspection, the inspector should ensure the extinguisher is in an area that is known to all employees in the work area and easy to access. The needle in the pressure gauge should be in the green, which means the extinguisher has appropriate pressure. Any physical damage or other apparent deficiencies that can be identified by visual analysis should be addressed. When in doubt, the extinguisher should be removed from service until it can be more thoroughly inspected by a fire safety professional.

OSHA also requires that fire extinguishers undergo an annual maintenance inspection, which is more involved than the monthly inspection. This is typically performed by a fire safety professional, and any deficient units will be refilled or replaced at this time.

Common extinguishing agents include water, carbon dioxide (CO_2), dry chemical, and various types of foam-like agents. Fire extinguisher technology advances have resulted in refinements to these basic extinguishing agents and the introduction of newer agents, some of which are hybrids of the basic types. When the time comes to conduct the annual maintenance inspection, an ASP may suggest to management the review of newer and more versatile extinguisher types for purchase instead of refilling older units and/or replacing deficient units.

There are five classes of fires, and not all fire extinguishers will be acceptable for use with all of them. An ASP should commit to memory the fire classes; extinguishers will identify the types of fires they are capable of dealing with by a letter designation representing a particular class of fire.

Class A (Ordinary Combustibles) fires include paper, wood, cloth, most trash, and also some plastics and other materials. Many of the combustible materials on work sites will fall into this category. Water, dry chemical, and foam are suitable for Class A fires. CO_2 extinguishers should not be used for these types of fires because the stream of pressurized carbon CO_2 could displace lighter combustibles, like paper, and inadvertently spread the fire.

Class B (Flammable Liquids and Gases) fires involve flammable liquids common to certain work sites such as gasoline and solvents and also some flammable gases like propane. CO_2, dry chemical, and some types of foam are acceptable extinguishing agents, although water is not an ideal agent for these types of fires. That is because some flammable liquids, when mixed with water, will still be flammable even though diluted. Additionally, the stream of water or foam may displace the flammable liquids and spread the fire. However, there are some types of foam-based extinguishers that employ an agent particularly suited to flammable liquids.

Class C (Electrical Equipment) fires should never be fought with any wet extinguishing agents for obvious reasons. CO_2 and dry chemical extinguishers are commonly employed for electrical fires.

Multipurpose dry chemical extinguishers are suitable for Class A, B, and C fires and are commonly found in the construction industry.

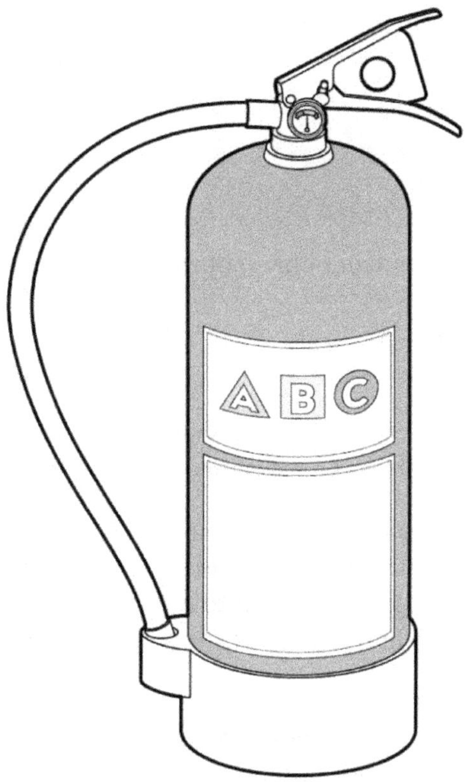

Dry chemical extinguishers are common to many workplaces because of their versatility, and an ASP can expect to see them on certain work sites, such as construction. Many are called ABC extinguishers because of their suitability with ordinary combustible, flammable liquids, and electrical fires; these letters will appear on the extinguisher. **Multipurpose dry chemical** is a phrase that refers to modern dry chemical extinguishing agents. On a modern construction site, for example, **dry chemical** and **multipurpose dry chemical** will refer to the same thing; much older dry chemical extinguishers were suitable for Class B and C fires only and have fallen out of use.

Class D (Combustible Metals) fires involve the alkali family of metals found in the periodic table, such as lithium, sodium, and potassium. These elements react violently with water and ignite when exposed to air. Only special-purpose "dry powder" extinguishing agents can be used on these fires, which should not be confused with dry chemical.

Class K (Oil and Fats) fires are those involving cooking oils, animal and vegetable fats, and other materials typically found in kitchen environments.

Fire Extinguisher Classes					
Class	**Material**	**Water**	**CO_2**	**Dry Chemical**	**Foam**
A	Ordinary combustibles	YES	NO	YES	YES
B	Flammable liquids and gases	NO	YES	YES	SOME
C	Electrical equipment	NO	YES	YES	NO
D	Combustible metals	Dry powder only (not the same as dry chemical)			
K	Oils and fats	Wet chemical			

The fire class letter assignments represent simple groupings and should not be confused as representing a greater or lesser hazard; Class A fires are not necessarily more severe than Class B, and vice versa.

When inside buildings during construction or other worksites activities, there must be no more than 100 feet of travel distance to the nearest extinguisher. In multistory buildings, there must be at least one on each floor, and of those, at least one must be adjacent to a stairway. If the building has a flammable storage room, then a fire extinguisher must be located within 10 feet of the door or entry to the storage room.

For outdoor construction or other worksites activities, an extinguisher must be located within 100 feet of open yard storage and within 50 feet of five or more gallons of flammable materials. For designated outdoor flammable storage areas, an extinguisher must be located at least 25 feet away, but no farther than 75 feet.

Sprinkler Types

Sprinkler types include: (1) dry pipe systems; (2) wet pipe systems; (3) deluge systems; (4) pre-action systems; (5) water spray systems; (6) foam water sprinkler systems.

Dry pipe systems are generally installed in locations that routinely experience freezing conditions. These are commonly installed in garages and buildings that are not heated. This system is not usually installed in a location unless the temperature reaches temperatures under 40°F. Once this system is activated, water is introduced into the system so that it may disperse into the distribution system.

A wet pipe system, a popular system, contains water within it at all times. When the sprinkler head opens after being activated by rising heat, the water is dispersed though the distribution system.

Deluge systems do not contain heat sensors or store water in pipes. A fire alarm system switches on the deluge valve and then water is discharged into the distribution system. The deluge valve must be manually reset to reset the whole system. It is often installed in locations with an increased potential for rapidly spreading fire.

Pre-action systems contain a water supply valve and functions like the dry pipe system. This valve is initiated by the detection of smoke or fire. The valve delivers water through the disbursement system, creating a wet pipe system. Subsequently, the heads are activated to release water onto the area. Generally, it is installed in locations where unintentional water discharge is unwanted so that valuable property does not get wet. Common installation locations include museums and IT operations room.

Water spray systems perform like a deluge system. However, the water is released in a configuration that is for particularly structured components and are generally designed for objects that have three dimensions. Deluge systems are different because the systems are structured to disperse over a location's floor area.

Foam-water sprinklers contain water that is mixed with foam with minimal expansion – foamy mixture. It is generally installed in locations that contain flammable liquids.

Segregation and Separation

Flammable Materials Storage

Segregating and separating materials and spaces are methods to prevent fires or minimize damage should a fire occur. OSHA sets forth regulations for flammable materials storage, including requirements for storage in relation to environments where heat and oxygen are present, as well as the materials used for storage. Finally, the pressure of storage tanks is also regulated. For example, most flammable liquids, such as oil or petroleum, are required to be stored in low-pressure containers made of concrete or steel and at a minimum of 20 feet away from heat or oxygen sources. Additionally, chemical storage guidelines dictate how to safely store various chemicals based on their individual qualities in order to minimize fire and explosion risks. This includes storing flammable and combustible liquids based on their flash points, the temperature point at which a fire or combustion could occur. Flammable and combustible liquids typically require storage areas that are stable in temperature and humidity and are not near any sources of heat; specific, controlled storage refrigerators or cabinetry are often required in laboratories or manufacturing facilities where such liquids are used.

Ventilation

Ventilation systems in buildings should promote positive pressure ventilation, which causes indoor air to be pulled outwards of the structure. This allows smoke and other noxious air particulates to leave an indoor space. It also reduces pressure in a space, which can reduce fire hazard risks. In buildings, especially in attached residential buildings, such as apartments or condominiums, units are separated by walls constructed of steel, concrete, or gypsum. This provides a barrier that protects adjacent units from a source of fire.

The different types of ventilation can be categorized as mechanical, natural, and mixed mode (hybrid). **Mechanical ventilation** involves exhaust fans, supply fans, or some combination of the two. Exhaust fans push air out of the building, while supply fans push air inside the building. **Natural ventilation** utilizes natural phenomena to passively move airflows between the inside and outside through strategically placed openings like windows and doors. **Mixed-mode ventilation** is most often used when natural ventilation systems cannot rely on environmental conditions to move airflows during some part of the day or for more prolonged periods. Some mixed-mode systems simultaneously run both mechanical and natural ventilation, while others alternate and moderate their usage at different times. Ventilation systems also commonly include features that target specific high-risk areas of contamination, such as a fume hood in a laboratory. Commercial buildings generally adhere to the Ventilation Rate Procedure to satisfy IAQ requirements, and this formula is based on the rate of ventilation that's required to meet industry and regulatory standards given the conditions at play, such as occupancy and usage type.

Many commercial buildings use **heating, ventilation, and air conditioning (HVAC) systems** for ventilation. Unlike other relatively simple ventilation systems, HVAC integrates ventilation alongside its heating and cooling functions. Ventilation in HVAC systems often has an air handler unit (AHU) to provide mechanical ventilation and has the ability to incorporate natural ventilation. Since HVAC systems require significant amounts of energy, some feature an energy recovery system for ventilation, such as a heat exchanger or enthalpy wheel, which transfers energy (heat) to incoming fresh air. In addition, HVAC systems usually monitor air filtration and cleaning through its clean air delivery rate, which depends on room size, filter type, and amount of airflow. Installation and maintenance of HVAC systems requires certification from a professional organization or government body, and the American Society of Heating, Refrigerating, and Air-Conditioning (ASHRAE) is widely recognized as the leading body of experts in the field.

Housekeeping

Housekeeping is a simple yet sometimes overlooked method of controlling fires. Materials that are disorganized, cluttered together, or improperly stored near heat sources can inadvertently serve as fuel sources for sparks that can lead to larger fires. Housekeeping practices that prevent or reduce fire risks include collecting, storing, and disposing of flammable wastes appropriately; even scrap wastes and shredded paper can serve as fuel sources, and clutter can lead to fire spread, as flames have more material to use as fuel. Grease and oil spills should be promptly cleaned, including in workplace breakrooms. Regular dusting also removes particulates that can lead to hazardous conditions, and wet wiping surfaces, such as fan blades, can attract combustible dusts that may otherwise disperse in the air. Tight nooks should regularly be cleaned with vacuums. Housekeeping should include timely and thorough inspection and maintenance of electrical and mechanical equipment since frayed wiring, faulty parts, and increased friction can increase fire risks. Flammable materials, including cleaning liquids, should be stored safely, away from sources of heat or pressure. Finally, preventative organizational policies, such as establishing designated smoking areas outdoors and away from equipment and flammable materials, are useful aspects of workplace housekeeping.

Emergency Response Management (ERM)

Emergency, Crisis, Disaster Response Planning

Plan for Emergencies

Organizations must develop a written emergency plan for potential emergency situations and disasters. Emergency plans must provide protection to employees, the public, and the environment. The best way to maintain health and safety in the event of an emergency is to conduct advanced planning according to project specification, conditions, and geographic area.

Planning for emergencies should be based on priority with respect to events most likely to occur and the potential severity of impact. With respect to naturally occurring events such as floods and earthquakes, organizations should conduct extensive research to determine what threats are most likely to occur within prospective project areas. To reduce organizational pressures, emergency and disaster services can be contracted to appropriate agencies. The following list includes manmade and natural events that organizations must consider when planning for emergencies:

- Fires and explosions
- Hurricanes and tornadoes
- Earthquakes
- Floods
- Disbursement of toxic chemicals in the atmosphere
- Terrorism and workplace violence

When writing emergency plans, organizations should seek cooperation with any neighboring entities and public organizations. Written plans must cover a multitude of elements including the following:

- Handbooks and other materials related to emergency procedures that are available to employees (e.g., facility layout, evacuation routes and assembly areas, locations of response/first aid equipment)

- List of potential/anticipated emergency situations and risk statements

- List of safety process elements such as incident investigations and hazard mitigation and controls

- Description of communication equipment and methods with contingencies

- Description of alarms and warnings

- List of cooperative measures and communications with local firefighters, law enforcement, utilities, and emergency medical services and first responders

- Description of performance elements such as drills (fire and evacuation) and training for fire extinguishers, medical and first-aid procedures, hazardous materials spill response, and rescue

A chain of command should be established for communication and reaction to a given emergency situation selected by the organization's advisory committee and/or emergency coordinator. Personnel should be selected based on their capacity to perform under stress, rather than their organizational title. The chain of command should be kept as small as practical for speed and efficiency in emergency situations.

Crisis

A crisis features three common elements: the element of surprise, the threat posed to the organization, and the time in which to act. Crisis management is the process by which an organization responds before, during, and after a crisis.

Emergency response equipment includes first aid kits, bandages, flashlights, eye wash stations, batteries, and other items necessary to respond to an emergency.

Emergencies on worksites can involve many types of materials and equipment. Examples include flammable liquids, corrosive chemicals, mechanized and motorized equipment, and even members of the general public, if they are not prevented from accessing the site.

Life Safety

In an emergency, the first priority is to safeguard personnel, equipment, and processes, in that order. Safeguarding the health and safety of personnel must always take precedence over equipment, tools, machinery, and all other assets.

Escape procedures and route assignments must be properly delineated; everyone on the site must know what to do and where to go during an emergency situation. In selecting an assembly (or muster) point during a site evacuation, an ASP should consider all factors and specific site conditions so that employees will not inadvertently be placed in further danger when proceeding to their assembly point.

At larger sites, designating multiple assembly points will allow employees to proceed to the nearest one when the alarm is sounded; this is especially advantageous when employees will be working at multiple places on the site during the workday. Assembly points should not automatically be fixed for the duration of the project; the assembly locations may need to change during the course of work and whenever any situations require a change, although this may not be necessary at smaller sites and for projects completing in a few days.

Whoever is selecting the assembly point should be familiar with the general geography of the site. This should include elevation, terrain features, landmarks, approximate distances, roads, and travel paths for vehicles and mobile equipment, along with any obstacles and barriers.

For sites involving the potential for a poisonous gas exposure like hydrogen sulfide (H_2S) (for example, oil field construction service and support operations), further considerations like wind conditions must be noted so that the assembly point is in the safest place possible.

Designated first-aid responders and all other medical response duties must also be addressed in the EAP. This is a formal and very important requirement that will ensure that only qualified personnel are delegated the authority to undertake these tasks; injuries have been inadvertently exacerbated by first-aid actions taken by well-intentioned but insufficiently trained first-aid responders.

Reporting procedures and designated contacts are also required to be in an EAP. Examples of common contacts include local fire and police departments, county emergency management personnel, regional poison control centers, and other relevant parties such as the hiring client and landowners.

An ASP will need to ensure that all emergency response provisions and responsibilities are communicated to employees. While OSHA does not require documentation of this, it is advised that employees acknowledge the receipt of this information, such as with an emergency orientation checklist or a simple sign-in sheet.

Disaster Preparedness

Disaster preparedness is a critical aspect of environmental safety. A disaster can be broadly characterized as an event or series of events with the potential to wreak havoc and/or seriously harm a person or location. First and foremost, disaster preparedness involves risk assessments. This analysis is context dependent, varying based on the location and the materials present. Generally speaking, risk assessments involve identifying all potential risks, evaluating the danger posed, and ranking the risks based on likelihood and severity. Following a risk assessment, disaster preparedness focuses on the promulgation of standards as to when a disaster should be declared, including the identification of preliminary events likely to precede a disaster. Lastly, disaster preparedness includes the creation of plans and protocols for responses aimed at avoiding, reducing, mitigating, and/or eliminating the disaster.

Disasters can be further classified based on whether they are manmade (anthropogenic) or natural. As the name implies, human actions are the source of **manmade disaster**. Examples of manmade disasters include biological disasters resulting from inadequate sanitation, chemical disasters based on a failure to contain hazardous materials, waste removal disasters, engineering disasters related to the collapse of infrastructure, and nuclear disasters arising from a malfunction at a nuclear power station. In contrast, **natural disasters** happen independently of human actions, though they can pose a similarly grave threat to people, facilities, and ecosystems. Manmade disasters are largely dependent on the types of materials and processes present on the worksite. For example, a laboratory running chemical tests would focus primarily on preparing strategies for preventing, containing, and mitigating chemical disasters.

Extreme weather events are the classic examples of natural disasters, and examples include blizzards, heat waves, hurricanes, tornadoes, wildfires, and hailstorms. Geological events such as earthquakes, avalanches, landslides, and sinkholes are another major cause of natural disasters. Location plays an outsized role in determining preparedness for natural disasters. For example, a laboratory located near a tectonic plate boundary in a tropical climate would prepare for earthquakes more than blizzards.

Disaster Recovery Plans

In the event of a crisis, an organization may face multiple challenges such as mitigating casualties, protecting property, and testing disciplinary protocol.

A **disaster recovery plan (DRP)** is a set of procedures that prepares for a disaster so that destructive effects are reduced, and essential data can be recovered. A DRP increases a firm's ability to recover from an unexpected, devastating incident. A DRP assists the organization in resuming normal business functions as quickly as possible. As information technology systems become more sophisticated and complex, solving critical organizational technology questions becomes more difficult. The ability of hackers and viruses to infiltrate these systems makes an effective organizational DRP more important than ever.

Technological increases have created viable scenarios for employers to offer alternative work locations. If a corporation offers alternative work locations, these employees are allowed to work from home or another off-site location rather than a traditional office space. Communication between organizations and its remote employees generally takes place through the Internet and phone calls. Alternative work locations can be helpful in disaster recovery because organizational data is decentralized and more difficult to corrupt entirely.

A procedure is a recognized and established way of accomplishing a desired goal. Procedures provide a plan of action for organizations in a time of vulnerability or crisis. The business continuity plan, disaster recovery plan, and any additional organizational policies should be studied and practiced frequently by employees.

Workplace Violence

Workplace violence is any act of physical violence, intimidation, threat, or verbal abuse that occurs in the workplace. This behavior is disruptive both physically and psychologically. Employees may demonstrate violent behavior as a result of a history of violence, a troubled upbringing, issues of substance use, and psychological illness. These conditions may foster violent behavior from an employee but do not make violent behavior inevitable. Workplace violence not only interrupts immediate employees, but can cause an organization to lose clients, suppliers, and advertisers. Furthermore, a firm can suffer devastating economic consequences as a result of negative publicity from incidents of workplace violence. Workplace violence attacks the foundation of trust and safety that all workplaces need to operate successfully.

Although an employer cannot completely eliminate the possibility of workplace violence, several steps can be taken to avoid these incidents. One example is a mental health program, such as an Employee Assistance Program (EAP), which provides employees the option to improve their psychological wellbeing. Additionally, offering company parties and functions in alcohol-free locations may reduce the likeliness of workplace violence. Violence may also be introduced in the workplace from the public. In areas with high crime rates, statistics show a higher probability of violence for employers who operate at night. Finally, organizations should establish and enforce a zero-tolerance policy for on-site weapons and acts of violence.

Shooting

Shooting at a workplace may be conducted by someone that is a known or unknown person. A person should reasonably protect their own life. First, a person should evacuate the area as soon as possible. Once in a safe location, the person should call 911. If a person is unable to evacuate to safety, then the next best option is to hide in a secluded location. The hiding place should be behind a locked barrier or blockade of heavy items. In this location, turn off lights and remain silent (including cell phones) until law enforcement personnel arrive and instruct otherwise. As a final resort, when their own life is in imminent danger, a person should confront the shooter. Finally, when approaching law enforcement or government agencies in this scenario, a person should approach the officials slowly with their hands in the air because the agencies may be unaware of the shooter's identity.

Companies should establish policies and procedures for active shooter situations. These measures, and related training, will hopefully decrease potential harm by providing personnel with a plan to follow in a moment of duress. Following the run, hide, or fight outline above may be helpful.

Bomb Threat

Bomb threats are oral or written threats by someone intending to use explosives in a particular location. Most threats are received by phone. A person should always take the allegation seriously. A receiver of a threat should stay calm and attempt to keep the caller on the phone. The receiver should ask questions to get as many details as possible about the threat, such as the person's location, identity, and other relevant matters. A receiver should take notice of other facts, such as the exact wording of the threat, voice identifiers of the caller, key words, and the number on the caller ID.

When the call is concluded, the person should immediately call a local law enforcement agency for assistance. It is imperative that the search for the explosive(s) be left to trained law enforcement. The receiver should evacuate the threatened location as soon as possible. Alternatively, if a package is received and suspected of being a bomb threat, then the receiver should not touch it. It should be left in its location, the area must be evacuated, and appropriate authorities must be contacted. The receiver of a bomb threat should be available for interviews from either management or local law enforcement.

Vandalism

Vandalism is conduct involving intentional damage or destruction of property. This may also include some acts of defacement or graffiti directed at the business entity or personnel of the entity. When vandalism is discovered, it is imperative that a person touch nothing. This preserves the scene for law enforcement personnel to gather facts and evidence of the offense. If vandalism is discovered and a person is unaware whether the offender is near, then the person should evacuate the area immediately. If evacuation is not an option, then the person should hide themselves behind a locked barrier or blockade of heavy items. Once in a safe location, a person should contact law enforcement personnel and request immediate assistance.

Verbal Threats

Verbal threats are oral statements that are made with the intent to harass or intimidate others within a workplace. Some may perceive such acts as a minor matter. However, depending on the statement made and the immediate circumstances, the verbal threat may escalate into the physical execution of the threat. A person should not respond to a verbal threat in a manner that would escalate the matter into physical harm. The best practice is to remain calm and not respond to the threat being made. Companies should establish a zero-tolerance policy for violence in the workplace and create site security plans to protect persons and property. Additionally, violence prevention programs should be implemented to raise awareness and knowledge of available resources and practices. Finally, businesses should encourage reporting of verbal threats so that such matters may be handled in a proper manner. Descriptive reports should be in writing and maintained to assist in developing safety policies and procedures for employees.

Industry Hygiene and Occupational Health

Recognizing hazards in work activities goes far beyond the most common types of hazards such as falls-from-height and confined-space hazards. Effective hazard assessment efforts must include an in-depth analysis of all operations. When work activities include processes such as welding, demolition, use of hand tools for grinding and cutting, and use of blasting agents, the resulting environment becomes a matter of **industrial hygiene**, or the protection of workers from health and safety issues.

Like other hazards, industrial hygiene issues must be recognized and assessed prior to the beginning of work activities. Subsequent control measures can then be used to mitigate risk to workers and the environment. Risk related to chemicals, toxins, and airborne contaminants is often based on the amount (or concentration) of the substance present and the duration of exposure. Common environmental and industrial hygiene issues are classed into three categories. These categories, along with the processes and substances associated with each, are as follows:

Chemical Hazards	Physical Hazards	Biological Hazards
Lead	Noise	Mold
Asbestos	Vibration	Bird droppings
Welding	Radiation	Bacteria and viruses
Solvents	Ergonomics	Poisonous plants
Corrosives	Heat and cold exposures	Animal/insect bites

Many types of evaluation methods are used to determine the extent of exposures to such substances. Methods include real-time monitoring or obtaining a sample of a substance in the work environment where results are compared to regulatory limits. Results can be used to determine if controls are necessary or if applied controls are adequate.

For many of these hazardous substances, **OSHA permissible exposure limits (PELs)** outline how organizations must comply with regulations regarding exposure to these substances and allow OSHA to enforce the limits of use and punish noncompliance to regulations. In contrast, the limits and values set by other organizations are recommendations and are unenforceable. The **National Institute for Occupational Safety and Health (NIOSH)** lists **recommended exposure limits (RELs)**. **Threshold limit values (TLVs)** are held by the **American Council of Governmental Industrial Hygienists**. Organizations can apply TLVs for a higher standard of worker protection, as they are often more stringent than the limits of OSHA.

Sources of Biological Hazards

Biological hazards are entities that can enter the human body and cause disease. Many hazards are not visible to the naked eye and can be present in many workplaces. They are most commonly found in workplaces that have many people working in them, workplaces that do not have high cleanliness standards, workplaces that are centered on waste storage or management, or in laboratories where biological agents are studied. Common sources of infection include contaminated food sources that are consumed, infected people who do not isolate themselves until the infection has cleared, infected animals, and poorly managed research settings. Biological hazards are least likely to be found in sterile environments, such as regulated cleanrooms. Viruses, bacteria, parasites, and fungi are common sources of biological hazards.

Viral Hazards

Viruses are pieces of genetic material within proteins and are not considered living. Viruses require a living host (e.g., a human, animal, or tree) and host cells in order to replicate and cause disease. Viruses attach and enter host cells with receptors that are found on the proteins that contain them. The type of receptor found on a virus's protein indicates what type of cell it can infect. Once a cell is infected, the virus uses the mechanisms and organelles found in the host cell to replicate and infect surrounding cells. This can occur through the lytic cycle, in which a virus replicates itself within a host cell until the host cell can no longer physically hold all of the new viral material. This causes the cell wall to burst and viral particles to overflow and enter adjacent cells.

Infection can also occur through the lysogenic cycle, where a virus causes genetic change within the host cell. As the host cell goes through its natural division or renewal process, any new copies will contain viral material. When enough cells are infected, the host may begin to experience symptoms of infection and may contain enough virus particles to infect other hosts. Viruses are most commonly spread through the air; respiratory droplets; direct contact with infected tissues or fluids; fomites; or other vectors, such as animals or parasites. Different viruses spread in different ways. For example, the common cold is caused by respiratory viruses that are spread through droplets. If an infected person coughs or sneezes, their droplets contain viral particles that can spread to another person. Many sexually transmitted diseases, such as herpes simplex, are caused by viruses that spread through contact with infected fluid or mucosal tissues.

Bacterial Hazards and Fungus

Bacteria are single-celled living organisms and considered to be the oldest organisms on earth. Fungi can be single-celled or multicellular organisms. Both bacteria and fungi can multiply rapidly within a human due to the warm, moist, and nutrient-rich environments that can be found in common sites of infection (e.g., throat, skin folds, mucus membranes). Bacteria multiply by dividing, while fungi multiply through spore production. While many bacteria and fungi can cause disease in humans, a number are highly beneficial and aid in human nutrition and digestion. However, some pathogenic bacteria and fungi are becoming resistant to traditional antibiotic and antifungal therapies. The bacteria and fungi that do not die from these medications ultimately reproduce to create more hearty and common strains that become extremely hazardous.

Parasitic Hazards

Finally, **parasites** are organisms that live off a host organism. They can cause disease from nutrient depletion or by proliferating to the point of causing sepsis in the host. Parasites are most commonly spread through physical contact with an infected person or animal or through contaminated waste, soil, food, or water.

Mold

Mold is made up of different types of fungi that grows in environments that have excess moisture. The more moisture in the area, the more likelihood of mold; mold requires moisture to grow. There is an increased likelihood of mold in environments that are high in humidity, have water damage, and are subject to temperatures greater than 70° Fahrenheit. Symptoms of mold exposure include nasal congestion, irritated eyes, and trouble breathing. If mold is discovered, it should be removed immediately under applicable state and federal practices and guidelines to minimize potential harm.

Mold growth may be prevented by keeping an area dry and immediately cleaning up spills or moisture accumulations that have occurred.

One type of mold is aspergillus, and it can cause aspergillosis. There are two types of aspergillosis: (1) allergic bronchopulmonary aspergillosis (ABPA) and (2) invasive aspergillosis. ABPA causes respiratory issues such as wheezing and coughing. However, invasive aspergillosis harms people with immune system deficiencies by damaging tissue within a body. Aspergillosis primarily affects farmer and grain worker occupations.

Protocol for Bloodborne Pathogen Control

Universal Precautions

OSHA's blood-borne pathogens standard requires that employees with a reasonable anticipation of exposure to blood or any other potentially infectious materials receive specific, annual training in blood-borne pathogens. Other potentially infectious materials include bodily fluid such as semen, vaginal secretions, pleural and pericardial fluids, amniotic fluid, and any bodily fluid that is visibly contaminated with blood. Since it is difficult to differentiate between bodily fluids that are potentially infectious and those that are not, many organizations will employ **universal precautions**, which treat all blood and fluids as if they are infectious.

Training topics must include identification and knowledge of blood-borne pathogen hazards, control methods, emergency actions, personal protective equipment (PPE), the hepatitis B vaccine, the employer's exposure control plan, and general housekeeping considerations. OSHA generally does not require training documentation, but blood-borne pathogens training is an exception, and training records must be retained for at least 3 years from the time of training.

OSHA's blood-borne pathogens requirements are found in 1910, subpart Z. While a General Industry Standard area, these requirements also apply to employees in the construction industry when there is a reasonable possibility of an exposure, such as designated first-aid responders, occupational health nurses, field medics, and any other personnel who are specifically designated to respond to injury events demanding immediate attention on a construction site. However, this does not include all employees on a construction site; dealing with blood should not be considered to be within the realm of normal experience for most construction employees, since no site should experience so many injuries of this severity that everyone needs to be trained in first aid and, consequently, in blood-borne pathogens.

Some construction project managers will propose sending all employees at the project site to first-aid training to ensure there will always be sufficient means to administer basic first aid in the event of a medical emergency. However, this is not necessarily the most ideal course of action in every circumstance.

When making a recommendation as to how many employees on a construction site will receive first-aid training and be designated as basic responders, an ASP should determine how many will be sufficient for a worst-case scenario while also taking into consideration the total cost of first-aid training and also blood-borne pathogens training, since that will also be required of all designated first-aid personnel. If a site will have thirty employees, and having four of them trained in basic first aid will more than cover any anticipated emergencies as well as provide for coverage when other first-aid responders are on vacations or otherwise away from the site, etc., then it may not be in the contractor's financial interests to send all thirty to first-aid along with blood-borne pathogens training.

An ASP must always consider the financial aspects of any decision involved in construction hazard control. First-aid and blood-borne pathogens training will involve course fees, paying employees while they are in training and not constructing, and also a much greater amount of training documentation and recordkeeping duties, which will take time and cost more. If the ASP has analyzed the situation and determined that having thirty employees trained in basic first aid will not make the site any safer than when four are trained, then this should be addressed with management so a well-informed decision is made. However, there are facilities and hiring clients who will require that all personnel on their property receive first-aid training. In that instance, the decision is already made.

Employers must make a hepatitis B vaccine available to all employees with job duties involving a reasonable exposure to blood and other potentially infectious materials, such as first-aid responders. OSHA does not require that employees receive the vaccine, but employers must make it available and must pay for it when employees elect to receive the vaccination.

An exposure control plan must be developed when one or more employees have an occupational exposure to potentially infectious materials. The plan must contain an exposure determination including a list of all job classifications involving a potential exposure, which in the construction industry will mostly apply to first-aid responders and other on-site medical personnel. The plan must be reviewed annually to determine if updates are necessary, and it must be made available to all employees for review.

Employers must provide adequate PPE for employees who will be in contact with potentially infectious materials. Examples include examination-style gloves and eye and face protection. Whichever the type of PPE, it must prevent potentially infectious materials from passing through and contacting employees' skin, eyes, mouths, and any mucous membranes, and also their clothing under the PPE.

PPE must be removed before leaving the work area and placed in a designated container for decontamination or disposal. Most of the PPE used for first-aid purposes in construction will be of the inexpensive and disposable type, so disposal is common in construction. Any contaminated equipment or tools must be decontaminated prior to being placed back into service.

Mutagens, Teratogens, and Carcinogens

Chemicals from which chronic exposure results in significant health conditions are classified as carcinogens, teratogens, or mutagens. Carcinogens are chemicals that are correlated with the development of cancers upon chronic exposure. Teratogens are chemicals that are correlated with embryonic or fetal malformations in pregnant women upon chronic exposure. Mutagens are chemicals that are correlated with genetic mutations upon chronic exposure. However, many chemicals can cause harmful clusters of symptoms upon chronic exposure that are not necessarily classified within these three groups but still cause significant health conditions that reduce a person's quality of life. Most of these symptom clusters fall under respiratory, liver, reproductive, or neurological medical conditions.

A **mutagen** is a chemical that has the ability to alter an organism's genetic material, which may lead to harm. This is known as a mutation (defective gene). Testing may be administered to identify whether the chemical has a mutagenic ability. This is concluded by an Ames test. A chemical's mutagenic trait has a high correlation of causing cancer abilities. The test uses a bacterial strain, *Salmonella typhimurium*, to test for mutations in a gene.

When a mutation occurs within the body, the body has the ability to identify and repair mutations. Sometimes, a mutation may not be repaired which can lead to cancer development. This occurs because the mutation interacts with a normal cell and transforms it into a tumor cell. A tumor cell may be: (1) benign; (2) premalignant; or (3) malignant. Benign tumor cells develop very slowly and are not cancerous. Premalignant tumor cells have the potential to develop into cancerous cells. Malignant tumor cells are developed cancerous cells.

A **teratogen** is any substance that may interfere with proper development of a fetus or embryo; it can be any substance that may cause a birth defect to a fetus or embryo during its development. Teratogens have the ability to affect the child's behavioral or emotional development. Additionally, it may also affect the child's intellectual ability.

A **carcinogen** is any substance or agent that is likely to cause cancer. Mutations in a cell's DNA may cause cancer. Introduction of these mutations may either be inherited or acquired through environmental exposures. These exposures may include such things as lifestyle choices and practices, workplace exposures, and pollution. Being exposed to a carcinogen does not mean that an individual will develop cancer; it means there is an increased likelihood of development. Multiple factors play a role in whether an individual will develop cancer from carcinogen exposure, such as duration intensity and the person's internal genetic structure. Carcinogen exposure may be limited by certain lifestyle changes, such as engaging in physical activity, reducing alcohol and tobacco consumption and usage, and avoiding lengthy periods of sun exposure. These practices will not completely eliminate the chance of developing cancer from a carcinogen, but they are small steps that an individual can take in an effort to protect themselves from cancer development. Again, individual genetics play a substantial factor.

Chemical Hazards

Sources of Chemical Hazards

Chemical hazards at the workplace are constantly changing as newer chemicals are introduced. Chemical hazards take the form of dust, fumes, gases, vapors, and liquids.

Chemicals take many routes to the body's interior, as they can be ingested, inhaled, or absorbed into and through the skin membrane. The daily habits of workers such as smoking and eating can introduce chemicals into the body when proper hygienic measures are ignored. Inhalation is the most common form of chemical interbody access.

Some chemical agents the ASP must be aware of include ozone, nitrogen oxide, carbon monoxide, chlorinated hydrocarbon solvents, mercury, fluoride, and cadmium.

The ASP must consult SDS in order to be aware of the properties of the chemicals in use. Consultation with the SDS for chemicals to be used will provide information for the material including physical properties (autoignition temperature, vapor pressure, and molecular weight), reactivity, toxicity, irritability, and flammability.

Assessment

Assessment of chemical hazards serves the purpose of assessing and comprehending the likely adverse effects of a chemical. Adverse effects may be environmental or harmful to people in nature. An assessment may be completed by: (1) hazard characterization; (2) exposure assessment; and (3) risk assessment.

Hazard characterization refers to being aware of the correlation between a chemical (and its amount) and the effects that it may have on persons or the environment – adverse effect probability. Exposure assessment refers to measuring or reasonably estimating an amount of time of being exposed to a chemical. Hazard is the potential to cause harm, but risk is the probability of harm resulting from a hazard. Risk characterization combines relevant hazard characterization and exposure assessment information to determine the probability and effect of a hazard risk. This probability also considers safety procedures in place and safety equipment that is available for use while handling a chemical. A hazard may not be reduced completely, but engaging in safety practices and conducting a risk assessment allows for risk to be reduced.

Control Strategies

Control strategies for chemical hazards are critical to worker safety as well as public health safety. Contaminants that spread within a work environment can also spread outside into the community. Controlling access to areas where hazards may be present is a critical and necessary first step for all work environments. Access should be limited to only those personnel who are required to work in these areas. **Access controls** may include badge entry requirements, secure and documented sign-in and sign-out processes, and automatic locking mechanisms that prevent entry by unauthorized personnel. Personnel with access should have the appropriate background and training to support safe work practices.

Personal protective equipment is another simple yet effective method of controlling exposure to hazards. Personal protective equipment may vary by work environment but generally includes methods to protect common entry points into the human body. Additionally, equipment may be used to mitigate the spread of hazardous material from inside the confines of a workplace to the larger community. Commonly utilized personal protective equipment includes gowns, gloves, shoe covers, eye goggles, masks, and hair nets. In work environments with especially hazardous environments, a hazardous materials suit may be utilized. These suits create an impenetrable, head-to-toe barrier for the worker against all microbes and other hazardous agents. Work environments should provide processes for utilizing protective equipment, such as contained changing and disposal areas.

All workers should also provide informed consent for their role. Standard operating procedures in a workplace should include how to incorporate controls into daily tasks, how to identify and report possible risks and exposures, and when and how to take action to preserve human health if exposure does occur.

Symptoms

Symptoms are effects that are experienced either physically or mentally that indicate harm resulting from a disease or condition. This expression shows that the body exposed to some harmful effect. Depending on the affected body portion, symptoms manifest in different expressions.

There are multiple factors that may affect the risk of a resulting symptom: (1) toxicity; (2) exposure area; (3) dose; (4) duration; (5) combined interactions and reactions; and (6) personal susceptibility. Toxicity refers to the level of toxicity of the chemical – the more toxic the chemical, the more likely it will be for symptoms to develop. Exposure area refers to how the chemical was introduced to the person, whether consumed, inhaled, absorbed through skin, or injected. Dose refers to the amount of chemical exposure a person experienced – higher amounts have an increased likelihood of symptom development. Duration refers to the amount of time the person experiences the chemical exposure – the longer the time, the increased severity of resulting harm. Combined interactions and reactions refer to chemicals having a greater resulting harm because of an additional chemical that was included in the exposure. Personal susceptibility refers to the likelihood that symptoms developed because of a person's individual characteristics – such as body size, age, sensitivities, etc.

Target Organs

Target organs refers to specific organs, or systems, which have been affected by a chemical and are causing a resulting symptom to that affected organ. Below are affected organs and corresponding symptoms that may have been caused by chemicals.

Head – vertigo (dizziness) or head pain

Eyes – irritated, runny, or red in color

Chest and lungs – trouble breathing, coughing, wheezing, or cancer affecting the lungs

Stomach – diarrhea, vomiting, or stomach pain

Skin – itching, redness, irritated blotchy skin, or cancer affecting the skin

Nervous system – impeded balance or coordination, irritability, spasms or contractions, or feelings of nervousness or anxiety

Reproductive system – (a) women: menstruation irregularities, miscarriage, harm to an egg or fetus and (b) men: damage to sperm or decreased amount of sperm production.

Exposure Limits

The ASP should monitor work activities to ensure chemicals are used and stored properly; all chemicals to be used for a specific task will be listed on the job safety analysis (JSA) with the SDS information included to determine proper usage and application for each specific task.

All chemicals will have applicable PEL, **recommended exposure limits (REL)**, and **time-weighted average (TWA)** concentration. The ASP will use available resources and information to determine the proper usage and application for each chemical.

Solid particles formed during the vaporization of a metal that cool to room temperature. The fumes produced by welding are an example of this. Airborne concentration of fumes is measured in **milligrams per cubic meter (mg/m^3)**, or in the case of lead, the measurement is in **micrograms per cubic meter ($\mu g/m^3$)**.

Gases that are typically suspended in a liquid state when at room temperature and pressure but go through an evaporation process; these are quantified in **parts per million** (for perspective, one part per million is roughly equivalent to four drops of liquid in a fifty-five-gallon drum).

Liquid droplets suspended in air due to their small size; measured in the same units as fumes.

Material molecules suspended in the air at room temperature; these are measured in parts per million in air concentration.

As with the silica particles discussed earlier, dust and/or fibers are created by abrasive or percussive work activities such as demolition, drilling, cutting, and sanding. Dust is measured by its airborne concentration in the same units as fumes and mist, although fibers are measured in **fiber per cubic centimeter (f/cc)**.

The route of entry to the body for chemicals is through ingestion, such as swallowing as a result of hand-to-mouth actions (eating or smoking in contaminated area), absorption (penetration through the skin), inhalation (primary route of entry for chemicals), and injection (entry through a skin puncture). Dust particles that are inhaled are generally ten microns or less (the diameter of a human hair is approximately 110 microns). Where dust particles are known to be ten microns or less, a respirator with **high-efficiency particulate air (HEPA)** filtration is recommended.

Ceiling limit is a chemical concentration limit that may cause a person harm. In practice, these limits should be avoided for safety purposes. These limits do not consider potential respiratory protections that a person may be wearing; the limit is based on a person without respiratory protection. The purpose of the limit is to deter substantial exposure – the limit must never be exceeded. It is the *ceiling* limit before exposure may cause harm to a person working with or near a hazardous chemical. Multiple government entities stress these limits should never be exceeded at any time. Otherwise, it may cause substantial harm to affected persons.

Immediately dangerous to life and health (IDLH) is a classification of hazards that have severe risk of harm to a person based on acute exposure. IDLH limits were developed so that employers would be able to provide sufficient respiratory protection equipment for employees that may be working with these types of hazards. If a worker's respirator were to fail, the limit refers to the hazard's chemical concentration in the air that an individual may be exposed to without suffering permanent harm. Under these circumstances, permanent harm refers to permanent damage to a person's health or severe eye or respiratory system damage that would hinder a person's ability to flee from the hazard.

Asbestos Exposure Hazards and Controls

Asbestos is a mineral used in different products primarily for its heat-resistant traits:

In its five forms, asbestos covers the spectrum of toxicity, flammability, irritant, and carcinogen:

- **Chrysotile**: The most abundant type of asbestos found in U.S. construction products; ubiquitous in floor- and wall-related materials and is white or gray in color.
- **Anthophyllite**: Used for its acid-resistant traits.
- **Amosite**: Brown/tan colored; applications are in insulation and exterior surface coating.
- **Crocidolite**: Blue in color, and applications are in filtration systems and insulation.
- **Actinolite and Tremolite**: Contained in mineralogical deposits such as vermiculite, which is used in insulation materials.

There are different types of asbestos, and all are thought to be the cause of certain short-term and chronic illnesses such as asbestosis and mesothelioma. Asbestos particles inhaled by a worker may lead to illnesses such as cancer. The ASP must know the exposure limits for asbestos, have a plan in place, and closely monitor progression of work, specifically in older building renovations, new builds collocated with older ones, and demolition projects.

Exposure limits for asbestos are as follows:

- Not to exceed 0.1 fibers per cubic centimeter (f/cc) of air during an eight-hour work shift
- Short-term exposure not to exceed more than 1 f/cc over half an hour
- Unless a negative exposure assessment at the worksite verifies a below-permissible exposure limit (PEL) environment, a daily protocol for testing with the requisite documentation is required.

Exposures to asbestos fibers have been linked to an array of lung diseases, including cancer. Owners of facilities must be aware of areas containing asbestos. Those performing work in the hazardous areas where asbestos exists must be informed prior to beginning work.

Lead Exposure Hazards and Controls

Lead is an elemental metal that is formed into numerous compounds. It is a dangerous substance and has caused, and continues to cause, both short-term and chronic illnesses in humans. Even minimal exposure can have a chronic and negative impact on a worker's health and capacity for employment.

Lead is known to affect both male and female reproductive systems with the potential for stillbirth or miscarriage for the mother. Surviving offspring have an increased chance of mental or behavioral disorders.

Lead enters the system by ingestion or inhalation, and a significant portion of this will invariably end up in the circulatory system. Lead is typically utilized in plumbing, paint, liner application, and conduit materials on the worksite; workers will find the highest potential for lead exposure during the renovation or demolition of older buildings. Lead standards may include metallic lead, organic lead soap, and organic lead compounds.

Maximum PEL for worker lead exposure is fifty micrograms of lead per cubic meter of air during an eight-hour period. Exposure must be proportionally reduced for any time spent that exceeds the eight-hour period. Established lead standards on a worksite may apply to any work activities involving potential lead exposure to a worker; some of these activities may include material installation, structural salvage or demolition, renovation, materials removal, and transport and storage. Employers are responsible for a lead exposure assessment prior to project commencement.

Utilized in paint and as an additive for steel, lead is an inhalation and ingestion hazard whether manifested as fumes (when it is volatilized) or dust (created by abrasive action). The OSHA-**permissible exposure limit** (PEL) for lead is 50 ug/m^3 during an eight-hour **time-weighted average** (TWA). The OSHA **Action Level** (AL) is 30 ug/m^3 per eight-hour TWA.

Silica Exposure Hazards and Controls

Silica, like asbestos, can cause serious and long-term lung ailments. Quartz materials are the primary source of crystalline silica and are abundant in a number of materials such as mortar, cement, granite, bricks, tile, and sand. Mitigating measures include an assessment of the work environment to determine the proper exposure limiting measures, which may include ventilation, worker position alternatives, tool modifications, controlling worker exposure with time on-site, and respiratory gear. As a worksite can be a fluid, shifting environment, it is essential the ASP constantly monitors the often-changing on-site work conditions.

Quartz is the most common form of silica and is contained in sand, clay, granite, gravel, and other types of rock; it is one of the more common minerals on the planet. Exposure to silica can occur in abrasion-related work such as sanding, scraping, grinding, drilling, or cutting. The extent of any detrimental health effects resulting from silica depends strictly on the size of the airborne particles ingested over the period of time the worker is exposed to the material. The most common illness associated with silica is silicosis; like hearing loss, it is preventable but not curable and quite serious, as it creates scar tissue in the lung that inhibits the extraction of oxygen from respiration.

The following are three classifications of silicosis. *Chronic silicosis* is defined by exposure over ten or more years. **Accelerated silicosis** is defined by increased exposure over a shorter time period (five to ten years). **Acute silicosis** is defined by a concentrated exposure manifested in a few weeks or up to five years. Beach sand is an example of very large silica particles that are mostly harmless simply because of their size.

Silica is an irritant, is classified as a carcinogenic, and can actually be flammable given sufficient airborne density and an ignition source. As particle size and density for airborne silicates are produced relative to the task being undertaken, the ASP must assess the circumstances and proper controls with which to mitigate the hazard, while the employer is accountable for providing such resources.

Routes of Entry

Chemicals enter the body through four main pathways: inhalation, ingestion, absorption, and injection. Chemicals can be found in various states of matter, such as gas, liquid, powders, mists, and solids; therefore, chemicals enter the body in different ways.

Chemical hazards can be mitigated through properly ventilating workspaces, such as having screened windows or vented hoods that lead outdoors, and using protective personal equipment that safeguards a worker's nose, throat, skin, and eyes. Additionally, establishing procedures that support proper storage, use, clean-up, and disposal of workplace chemicals can help reduce risk of exposure. For example, banning food or drink in laboratories where chemicals are used can reduce the risk of inadvertent ingestion. Specialized equipment, such as eye washing stations, can help mitigate or reduce the effects of chemical contact should it occur. Finally, proper handling and disposal of sharps and broken objects, such as glass equipment, can reduce the risk of chemical injections.

Inhalation

Inhalation occurs when chemicals enter the body through the nasal or oral cavity and make their way into the lungs. Chemical inhalation can occur when gases or fine particulates are released into a confined, poorly ventilated environment. Once chemicals are in the lungs, they are easily able to pass into the bloodstream as blood oxygenation occurs. Initial physical symptoms may include nasal or throat irritation and coughing. As chemicals enter the bloodstream, symptoms can include headache, dizziness, fatigue, and disorientation; without intervention, the person may become unconscious and collapse. Some chemical agents, such as carbon monoxide, are so noxious that they can cause death by poisoning within minutes. However, most instances of chemical inhalation can be resolved quickly if they are acknowledged early enough. By simply introducing fresh air and ventilation, the human body is typically able to clear out most noxious agents through the respiration process. However, some lingering effects, such as a sore throat, can take several days to resolve depending on the type of chemical that was inhaled. More serious cases may require concentrated oxygen therapy.

Ingestion

Chemical ingestion occurs when a chemical is consumed and enters the stomach, therefore affecting gastrointestinal organs and possibly entering the bloodstream through this pathway. Symptoms of chemical ingestion can include an unpleasant or unnatural taste in the mouth, stomach pain, vomiting, and general malaise.

Absorption

Chemical absorption can occur when a chemical comes in contact with the eyes or skin. Chemicals enter the bloodstream more quickly through this type of contact, as compared with inhalation or ingestion. Eye and skin damage can quickly occur and cause long-lasting effects, such as affected vision, burns, or scarring.

Injection

Chemical injection is another entryway through which chemicals can quickly enter the bloodstream. Chemical injection occurs when a sharp object that contains chemicals pierces the skin and inadvertently deposits chemicals directly into the bloodstream.

Acute and Chronic Exposures

Acute and chronic exposures to chemicals can cause different effects to the human body. Both acute and chronic exposure to chemicals are more dangerous to vulnerable populations, such as young children and obese, immunosuppressed, or elderly citizens. These groups have less developed immune and cardiovascular systems and therefore cannot quickly clear most chemical toxins. Additionally, many chemical toxins can build up in fat stores. The Occupational Health and Safety Administration (OSHA) has compiled a list of Particularly Hazardous Substances (PHS's) that outlines best handling practices for many of the carcinogenic, teratogenic, mutagenic, and other disease-causing chemicals (e.g., cyanide, arsenic, chlorine). These practices outline who should truly not work with such substances (e.g., pregnant workers); guidelines for reducing contact for all workers; and methods for storing, handling, and disposing of such chemicals.

Acute exposures refer to short-term interactions with a chemical, and they can cause side effects or symptoms that last for a short period of time immediately after exposure takes place. They typically seem minor and appear to resolve quickly. Acute short-term effects of chemical exposure often include symptoms such as irritated eyes, nose, or throat; sneezing; coughing; or skin tingling. Effects often go away when the exposure is removed. However, acute exposures to harsh chemicals, such as some acids, can cause severe symptoms quickly, such as burns. These are acute exposures from which effects take a long time to resolve.

Chronic exposures take place consistently over an extended period of time. Some mild symptoms may appear after each exposure. Generally, the cumulative result of exposure results in a significant, major condition. For example, cigarette smokers are exposed to various harmful chemical agents in each cigarette. After smoking a single cigarette, a person may notice mild effects from the chemicals, such as feeling alert or relaxed, a sore throat, or a funny taste on their tongue. After habitual cigarette smoking and chronic exposure to these chemicals, the smoker may experience more severe conditions, such as lung cancer, hypertension, or limited taste.

Additive Effect

Additive effect is a term used to describe the effect of two chemicals that are combined. Acting independently, two chemicals have caused a proportional effect. When combined, the effect is also proportional. That is, although an effect is increased due to the combination, the increased effect is proportional to the sum of the independent effects.

For example, consider consuming alcohol and a depressant. Both reduce stimulation of the body. When combined, the effect is proportional to the sum of the two acting independently. Commonly, mathematical expressions have been used to illustrate the concept of the effects of exposure.

Additive effect: 2 + 2 equal to 4.

Synergistic Effect

Synergistic effect is another term used to describe the effect of two chemicals that are combined. When these chemicals are independent, the effects may be nominal. However, when combined, it has a substantial affect that is greater than the sum of each independent effect.

For example, consider the harm that smoking and UV radiation cause to skin. Smoking causes harm to collagen and elastin in skin and may cause premature aging of the skin. UV radiation may also cause premature aging of the skin. However, combining smoke and UV radiation causes substantially more harm when the effects are combined on the skin.

Synergistic effect: 2+2 is greater than 4.

Antagonistic Effect

Antagonistic effect is another term used to describe the effect of two chemicals that are combined. However, this is different from additive effect and synergistic effect. Rather than combining the exposures to have an increased effect, the effect is impeded by the combination of the two. Either one or both exposures cause a decrease in the other exposure, and the combined effect is lower than if the exposures were acting independently.

For example, consider poison and an antidote. Poison causes substantial harm by damaging tissue. An antidote is designed to reduce the effects of poison. When combined, the antidote hinders the effects of the poison and prevents further damage to a poisoned organism.

Antagonistic effect: 2+2 is less than 4.

Potentiation Effect

Potentiation effect is the final term used to describe the effect of two chemicals that are combined. However, this differs from the above effects because this is used to describe a combination that occurs when one of the exposures has no effect. One exposure does not influence an organism, but the other does. When combined, like synergistic effect, the effect is greater than if the two effects were acting independently on an organism. The exposure that has no influence independently influences the other chemical and exaggerates its effect on an organism.

Potentiation effect: 0+2 is greater than 2.

Noise

Noise in the worksite can affect the worker on a short- or long-term basis; although noise-related damage may be prevented, it cannot be repaired once it has occurred. Hearing damage interferes with the reception of high-frequency sounds and will be detrimental to the worker's ability to communicate in and out of the worksite. As a result of the known effects of a high-noise environment, measures have been taken to mitigate associated hearing damage. To this purpose, the following maximum exposure limits apply: It is recommended the human ear should not be subject to decibel counts exceeding ninety dBA (OSHA) over an eight-hour period. Shorter exposure times at higher decibel counts can cause instant or long-term damage.

To reduce hearing hazards in the workplace, a noise hazard control process can be implemented with the following steps:

- Reduce noise through use of quieter equipment.
- If feasible, move and arrange worksite equipment away from workers so they may be isolated from noise-producing machinery or tools.
- Engineering controls may be used to provide sound barriers to isolate workers and reduce noise hazards.
- Hearing protection should be worn.

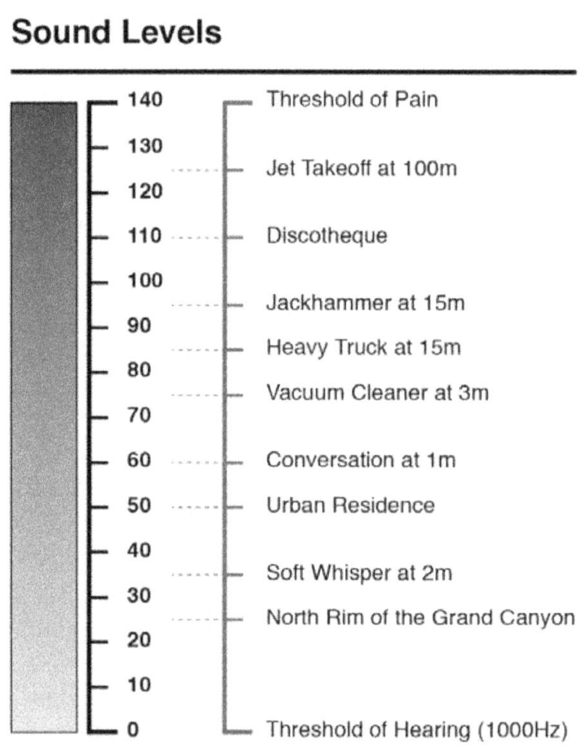

Noise is also considered a stress-inducing agent to be controlled with prescribed methods. OSHA mandates a 90 dBA limit during an eight-hour workday, although NIOSH recommends a lower limit of 85 dBA for the same amount of time. NIOSH does not decide the regulations; instead, they carry out specific workplace studies and make the workplace recommendations resultant from these studies to OSHA. The ASP must be aware of worksite noise levels and must know the procedures and means to reduce worksite levels to ALARP. Known control methods will be utilized to reduce the risk and deal with the hazard specific to its location and nature. The ASP, after identifying and assessing the noise hazard, will work with other safety and operational personnel to decide which control measure (machine relocation, ear plugs, noise barrier, HSE posted signage) will reduce the hazard to a manageable level (for decibel level and time of exposure).

The ASP must also know that OSHA states that if the ambient noise increases by 5 dBA or more, the exposure time for a worker must be cut in half; NIOSH recommends doing the same, albeit based on a 3 dBA increase in ambient noise.

Radiation

Radiation is energy moving through space manifested as electromagnetic waves or as subatomic particles. Ionizing and non-ionizing are the two primary forms of radiation.

Ionizing radiation is the product of electrically charged particles (ions) colliding with and reacting to materials. Ionizing radiation occurs naturally (sun, space) but is also found in things humans build and utilize (x-rays, nuclear power plants). X-rays, gamma rays, alpha particles, beta particles, and neutrons are all sourced from ionizing radiation.

Non-ionizing radiation cannot energize atoms, but it can create the movement and friction necessary to create heat. Non-ionizing radiation comes in the form of lasers, **radio frequency (R/F)**, **ultraviolet (UV)** lights, and R/F microwave signals.

The ASP should control measures to reduce the associated hazards of exposure, focus to reduce the time of exposure, increase the distance from the radiation source, and implement appropriate shielding of radiation sources.

Ionizing and non-ionizing are two types of radiation potentially present on a work site. Both are forms of energy differentiated by the amount of kinetic energy emitted. Ionizing radiation is manifested in particles or waves with sufficient energy to displace electrons from atoms. In its quest for stabilization, a radioactive atom's nucleus will emit high-energy waves and particles during the process of radioactive decay. Decay can be natural (radon gas is produced by the natural decay of radioisotopes) or man-made (splitting the atom at a nuclear reactor). Alpha particles, beta particles, and gamma rays are the primary radioactive materials produced by the decaying process.

Alpha Particles

Positively charged, high-energy particles comprising two neutrons and two protons. They are the product of decay in heavy radioactive elements, and although highly energetic, they move through the air slowly, as they are heavy in mass. These particles are more dangerous internally (swallowed or entering the bloodstream) than externally, as they will likely not penetrate the skin and will not penetrate through a sheet of paper.

Beta Particles

Like alpha particles, these can move through the air for considerable distances and are naturally occurring or man-made. These fast-moving electrons can, in fact, be stopped by clothing layers or, specifically, thick solid substances such as aluminum, and are most harmful when introduced internally.

Gamma Rays

X-rays are a form of gamma rays and are the primary form of man-made radiation exposure. Gamma rays are clusters of energy of a weightless nature (photons) and are frequently transported with the alpha and beta particles that are emitted from an atom's nucleus. Most gamma rays directed at the human body will simply pass through, but a few of these particles will typically be absorbed. Gamma rays are much more penetrative and will travel through several feet of concrete or several inches of lead before they are stopped.

Exposure and Control

The average American may be exposed to as much as 360 millirems per year (a **millirem** is one thousandth of a rem unit that measures exposure to radiation) either from external or internal absorption or ingestion. Radiation introduces unspent energy into body tissue, which might damage or kill cells or be causal in the development of cell abnormality.

Acute or chronic exposure to radiation consequently varies, and effects can be seen immediately or after long time periods; damage to the body will depend on radioactive energy absorbed by the body, time of exposure, dosage rate, and the parts of the body that were exposed.

Exposure to a single, large dose of radiation or concentrated exposure to a smaller amount in a brief time period. Exposure to small amounts of radiation over a longer time period.

The control methods for ionizing radiation include shielding, time, and distance from ionizing radioactive sources.

Non-ionizing radiation does not possess the energy to remove electrons (to ionize) from atoms or molecules, hence its difference from ionizing radiation. Ultraviolet (UV) (the sun), lasers, microwaves, and infrared are typical examples of non-ionizing radiation.

Control methods involve measures taken to protect workers from the effects of non-ionizing radiation (sun exposure, welding, heat lamps, and microwaves).

The control hierarchy must be employed by the ASP to assess and assign the control to a specific task where ionizing radiation or non-ionizing radiation presents potential exposure risks. This can be PPE, barriers, distance from source, or any controls deemed appropriate to the task. The ASP will assess the identified hazards for each task. Mindful of the fact that every task is indeed unique, the ASP monitors the development/application of task-specific job safety analysis (JSA) and the ongoing vigilance for associated workplace hazards.

Heat and Cold Stress

Workers can be exposed to extreme cold or extreme heat when working outdoors or in indoor sites without climate control. Working in extreme temperatures presents many hazards to workers' health such as tissue damage, frostbite, heat exhaustion, and heat stroke.

Extreme Heat

Workers exposed to extreme heat are at risk of heat-related illness such as heat rash, cramps, exhaustion, or heat stroke. Exposure to direct sunlight can cause sunburn, which can be severe enough to cause blistering and permanent tissue damage. Heat rash occurs when the skin becomes inflamed because sweat glands are blocked.

In addition to skin damage, if the body continues to be exposed and lose fluid and electrolytes, heat exhaustion can set in and cause muscle cramps, sweating, rapid pulse, headache, and vomiting.

To prevent heat-related illness, workers should wear lightweight, sweat-wicking clothing and hats to protect them from sun exposure and sunburn. Wearing cooling vests can also help eliminate overheating. They should drink water regularly to prevent dehydration.

Extreme Cold

Working in extreme cold exposes workers to the risk of tissue damage and cold-induced illnesses such as hypothermia and frostbite. When the human body is exposed to extreme cold, it becomes stressed, and blood is rushed from the extremities to keep the person's core temperature warm. Over prolonged periods of time or in extremely cold temperatures, this can cause frostbite, which can lead to permanent tissue damage. If body temperature drops below 95 degrees, hypothermia can occur. Severe hypothermia can cause shivering, dizziness, loss of consciousness, and, if not treated in time, death.

To prevent frostbite and hypothermia, workers should always keep their heads and hands covered and wear insulating socks and clothing to prevent heat loss. In cold temperatures accompanied by high winds, they should protect their eyes and faces. Workers should drink plenty of water to prevent dehydration.

Conducting an Exposure Assessment

Risk assessments for hazardous exposures should be conducted in all workplaces, and especially stringent assessments should take place in high-risk settings, such as microbiology laboratories. Assessments should be performed at regular intervals, with reminders for adherence occurring daily. Formal exposure assessments may be performed by trained, third-party risk management professionals; however, minimizing risk is the job of all stakeholders in the work environment. Administration, laboratory workers, researchers, janitorial staff, and anybody who may be impacted should feel comfortable conducting and communicating the results of regular, internal assessments.

A thorough exposure assessment should begin with documenting all parts of a workflow. This includes how the work process begins, who is involved, and what steps take place from start to completion of the task. The assessor should identify what hazards and exposure risks occur at each step. Once a list of exposure risks is compiled, each risk should be evaluated for likelihood of occurring and how severe the consequences might be. For example, if a laboratory worker fails to don all personal protective equipment before working with microbes, the risk of exposure is high. However, if the microbe is not known to cause disease in humans, the consequences are much lower than if it is a highly pathological microbe. Some exposures will be too dangerous to continue work; the workflow should be changed, modified, or eliminated before work resumes. For other exposures, the work environment may have controls in place to support human health and safety.

An important reason to regularly assess exposure and risk is because both are constantly changing. For example, a researcher who studies infectious parasites in a lab may one day learn that he has cancer. During treatment, the researcher will be immunocompromised. This may change the level of risk that he feels comfortable tolerating while working with infectious parasites, even if safety protocols are in place. Changes in supplies, materials, staffing, and external threats can also change the risk to exposures in a workplace. For example, consider if a large portion of janitorial staff became ill and stayed home from work. Without substitute workers in place, general cleaning processes may not get accomplished. This could expose other workers to pathogens that may not have been present otherwise. During the COVID-19 pandemic, many healthcare workers regularly faced shortages of personal protective equipment. This shifted the risk of exposure each day due to an unanticipated disruption in their normal supply. These examples show why workplace biohazard exposure should be assessed regularly, with informal discussion for work procedures taking place on a daily basis.

Environmental Management

Environmental Hazards Awareness

Environmental hazards play a major role in disaster preparedness and risk assessments because these hazards are generally the principal source of harm during disaster scenarios. Therefore, all personnel must be aware of all the potential environmental hazards at play, which varies based on the context and circumstances. Some of the most common environmental hazards include biological hazards, chemical hazards, waste, and vermin.

Biological Hazards and Mold

Biological hazards are living organisms that pose a danger, especially in terms of threatening people's health. Examples of biological hazards include bacteria, microbes, viruses, and natural toxins. The United States Centers for Disease Control and Prevention (CDC) categorizes disease-based biological hazards into four categories, with Level 1 having the lowest risk (for example, chickenpox) and Level 4 posing an extreme risk (for example, Ebola). Mold is one of the most commonly found biological hazards. While the overwhelming majority of mold is relatively harmless, some types of mold can trigger allergic reactions, mycosis (fungal infection), and potentially deadly poisoning. Molds primarily cause harm when people contact, inhale, or ingest mold spores and metabolites, like mycotoxins. Flood damage and poor circulation are the leading causes of mold because wet, humid, and stagnant air facilitates growth. To eliminate mold, the source of moisture must be eliminated before remediation can be effective.

Chemical Hazards

Chemical hazards take a wide variety of forms, ranging from flammable liquids (for example, diesel fuel) to irritants (for example, hydrochloric acid). Some other broad categories of chemical hazards include compressed gases, oxidizers, peroxides, carcinogens, and corrosives. The most common method of exposure is inhalation, but other routes include ingestion, injection, and contact with the skin and eyes. The primary method of first aid for chemical hazard exposure is to rinse the affected area with water, though some chemical agents require emergency medical attention and/or pose long-term health risks.

Hazardous Waste

Hazardous waste appears in solid, liquid, and gaseous forms, and it is defined by the presence of at least one of the following characteristics: corrosivity, ignitability, reactivity, and/or toxicity. As a result of the danger it poses to the public health and environment, the treatment, transportation, storage, and disposal of hazardous waste is heavily regulated. The Resource Conservation and Recovery Act (RCRA) establishes federal regulations and issues permits for handling hazardous waste, and states have similarly restrictive regulations and permit requirements.

Vermin

Vermin are animals that pose a health risk to people and threaten the sanitation of facilities. Examples include mice, rats, snakes, cockroaches, flies, mosquitoes, and termites. The primary method of eliminating vermin is exclusion, including the denial of access, elimination of resources, and use of repellants. If already present, vermin must be exterminated to prevent contamination.

Water

Stormwater, wastewater, and water usage are heavily regulated due to their potential environmental harm. The source of regulations and authority for permits is mostly based in the Clean Water Act (1972), Safe Drinking Water Act (1974), and the Resource Conservation and Recovery Act (1976).

Water conservation has become a top priority for governments due to increased water scarcity. In order to manage fresh water and safeguard the hydrosphere, federal and state governments have enacted numerous regulations and permit requirements for water usage, including limitations on the amount of water used and the modification of surface water areas.

Stormwater

Stormwater is created when rainwater travels back into the surface water and groundwater while carrying materials found on the ground, including but not limited to chemicals, grease, metals, nutrients, oil, and pathogens. The negative consequences of stormwater include water contamination, fish kills, and algae blooms. As a result, the EPA and state agencies regulate materials that can be left on the ground, especially on paved surfaces. In addition, permits are sometimes contingent on facilities following stormwater best management practices, such as limitations on contaminant discharge and the installation of filtration systems in storm drains.

Wastewater

Wastewater is any water contaminated as a result of human activity, including agricultural, commercial, and domestic usages. The chemical and biological contaminants in wastewater are varied, and examples include bacteria, emulsions, heavy metals, microplastics, pharmaceuticals, and thermal pollution. Depending on the location, wastewater can either have its own sewer system or be carried alongside stormwater and sewage; however, wastewater treatment plants must treat wastewater before it can be safely discharged. The National Pollutant Discharge System (NPDS) provides permits and establishes limits for discharging industrial and commercial contaminants. Some industries face additional permit requirements and regulations, including mining, oil and gas, animal husbandry, and agriculture.

Best Practices

Water pollution refers to the contamination of bodies of water, such as oceans, lakes, rivers, and groundwater. Given the importance of water to human civilization and ecosystems, water pollution constitutes a virulent danger, accounting for more death and disease than any other single factor. Therefore, safety professionals must seek to prevent both point source and non–point source water pollution. Point source water pollution is the direct contamination from a distinct and identifiable pollution generator, while non–point source pollution refers to an accumulation of pollutants, as in industrial or agricultural runoff pollution. The sheer variety of potential contaminants similarly poses a complex challenge for safety professionals. Along with chemicals and pathogens, contaminants can also include sediments, minerals, nutrients, and thermal changes. Consequently, safety professionals must have a strong grasp on the type of pollutants produced at the facility and all relevant regulations related to the discharge of those pollutants, including federal, state, tribal, and local regulations.

In the vast majority of cases, facilities send wastewater to sewage treatment plants, but if a facility is located in an area without a treatment plant, then a septic tank is needed to treat the water before it's released into the soil. If a facility produces wastewater with an especially high concentration of contaminants, such as oil or ammonia, then regulations might require the facility to pre-treat the wastewater before discharging it. For large amounts of thermal pollution, engineering controls include cooling ponds and cooling towers. Specially designed containers and spill prevention plans can prevent the release off of toxins in runoffs. Furthermore, industrial runoff can be controlled through the installation of infiltration basins, retention basins, and bioretention systems to prevent flooding, reduce erosion, and remove contaminants.

Air

Air Quality

The federal government and states have sought to protect air quality in order to increase visibility, safeguard agricultural production, prevent building damage, reduce acid rain, reverse ozone depletion, and protect vulnerable populations, including the elderly, children, and people with respiratory issues. Following its amendment and expansion in 1970, the Clean Air Act has become one of the most comprehensive pieces of air quality legislation in the world, and it empowers the EPA to regulate all forms of air pollution in coordination with state and local officials. In addition, Title V of the Clean Air Act established a national permitting program for commercial sources of air pollution. Consequently, businesses must measure and report their emissions to state and local environmental agencies as well as present and execute plans to minimize those emissions. Amongst other regulations, the EPA places especially strict criteria on the following pollutants: carbon monoxide, lead, nitrogen dioxide, ozone, particulate matter, fine particulate matter, and sulfur dioxide. More recently, the EPA has begun regulating the emission of greenhouse gases, such as carbon dioxide and methane.

The EPA, states, and local authorities also impose numerous regulations and permitting requirements to protect indoor air quality (IAQ). Depending on the jurisdiction, there are limitations on the amount of microbes, particulates, organic compounds, radon, and carbon monoxide that can be in the air. Carbon monoxide, radon, and asbestos fibers typically face the heaviest restrictions due to the associated risk of life-threatening harm. As a result, construction and building permits typically must meet the standards for IAQ. Issues with IAQ are most commonly resolved by filtration, ventilation, and/or removing the source of contamination.

Best Practices

Air pollution needs to be strictly limited to satisfy government regulations and avoid costly fines, so the safety professional must be familiar with air quality engineering controls. Although the specific type of engineering controls depends on the specific type and amount of emissions produced at a facility, the limitation of air pollution follows a common framework. First, safety professionals develop models to better understand the source and type of air pollution. This evaluation generally involves identifying the presence of potential contaminants, conducting an analysis of air patterns, and developing statistical models. Second, safety professionals must identify all applicable federal, state, and local regulations, and then they must determine whether the facilities are in compliance. These assessments typically include a review of permits, the Clean Air Act, state administrative rules, and/or industry best practices and compliance standards. Third, safety professionals design and execute solutions to keep or bring the facility into compliance. Fourth, the solution must be consistently monitored for effectiveness and adjusted accordingly.

The specific type of solution will vary based on circumstances, but there are a number of frequently used control mechanisms. For the control of particulates, control mechanisms include cyclones, wet scrubbers, electrostatic precipitators, and baghouse filters. Cyclones create a spiral airstream that forces large particulates to hit the outer wall, which are then collected at the bottom of the machine. Wet scrubbers spray miniature droplets of water or a liquid solution to remove particulates from an airstream.

Electrostatic precipitators use electrical charges and an electric field to trap particles. Baghouse filters suck air through a fabric-filter to remove particulates. If the emission is a gas, absorption and adsorption present potentially effective solutions. Absorption involves the capture of atoms, ions, or molecules in a solid or liquid, and the object of absorption becomes part of the solid or liquid's volume. In contrast, adsorption refers to trapping atoms, ions, or molecules to the surface of a solid or liquid. Activated carbon is the most frequently used adsorbent solid, and common methods of absorption include flue gas desulfurization, wet scrubbers, and packed scrubbers. Lastly, incineration can be used to convert volatile organic compounds (VOCs) and gaseous hydrocarbons into less harmful compounds in an afterburner.

Land and Conservation

Conservation and sustainability are critical aspects of protecting the environment and land from harm. **Conservation** involves preventing the wasteful consumption of resources and protecting ecosystems, including the maintenance of wilderness, restoration of habitats, and protection of biological diversity. **Sustainability** is a related concept, and it aims to create a balance between human civilization and the natural world to ensure the long-term viability of both. The principles of conservation and sustainability have shaped commercial regulations on how businesses dispose of solid waste and participate in recycling programs.

Land pollution primarily refers to soil contamination. When debris, chemicals, and waste are left on land, they eventually enter the soil, rendering serious harm to the environment. Land pollution can result in the eradication of entire ecosystems and pose severe dangers to human health, especially as a result of chronic exposure to heavy metals and carcinogenic chemicals. Examples of land pollutants include lead, herbicides, pesticides, petroleum, petrochemicals, sewage, and electronic waste. Therefore, safety professionals must adhere to regulations regarding waste disposal to ensure that materials aren't intentionally dumped or accidentally discharged onto the land during transportation or production.

Like prevention of air pollution and water pollution, prevention of land pollution is a much more sustainable and cost-effective practice than remediation. Strict protocols must be established for the arrival, use, and disposal of materials, especially for materials with the potential to have an adverse effect on the land. In particular, transportation containers must be properly sealed, and hazardous materials must be stored in appropriate areas and sheltered from harsh weather conditions. Furthermore, facilities must monitor workplace practices to enforce prohibitions against environmental dumping. Prevention measures can be tested through soil contamination monitoring tools, like a mass spectrometer and gas chromatograph. When a problem is identified, there are a number of engineering controls to mitigate the issue. Soil can be extracted and sent to a waste facility, or electromechanical and aeration systems can be used to break down contaminants in the soil. In the case of heavy metals, plants (phytoremediation) and fungi (mycoremediation) can be used to extract and/or accumulate the contaminant.

Solid Waste

Solid waste is colloquially referred to as trash or garbage, and the EPA has established a hierarchy of best practices for its disposal. Reduction and reuse are considered the most sustainable practices because they limit the consumption of natural resources as well as the generation of solid waste. Recycling and composting are considered to be the next most eco-friendly practices since these programs repurpose materials and organic compounds. Energy recovery is the next best practice because it transforms solid waste into energy, for example, through incineration. The least sustainable practice is disposal in landfills or garbage dumps due to their highly inefficient use of land and the transfer of contaminants into groundwater. Most jurisdictions require the separation of some forms of solid waste for treatment before disposal, such as hazardous waste, toxic waste, biomedical waste, and electronic waste. Depending on local ordinances, businesses are frequently barred from government-run waste removal and recycling programs and must make private arrangements with a certified company.

Recycling

Recycling refers to the conversion of waste into new materials. While recycling is relatively sustainable and supports conservation, it is considered an inferior option to reduce and reuse because significant energy is lost during the conversion of materials into new products. Some examples of highly recyclable materials include asphalt, concrete, glass, copper, iron, steel, paper, cardboard, and newspaper. Some forms of plastic are also highly recyclable, such as #1 PET plastic, and #2 HPE plastic, but in general, plastic constitutes a leading cause of pollution. In addition, plastic significantly strains landfills due to its lack of biodegradability. Although the EPA regulates landfills and establishes recycling objectives, recycling programs vary widely based on state and local legislation.

Sustainability

Sustainability refers to eco-friendly policies, and it has become an increasingly important aspect of workplace administration in recent years for three reasons. First, sustainability is strongly related to workplace safety because many materials are harmful to both the environment and humans. For example, many laboratories have increasingly pursued green chemistry, meaning the reduced use and production of hazardous materials, to mitigate risk and limit the deleterious effects of pollution. Second, sustainable policies and practices are often a component of regulatory frameworks. For example, sustainable practices in production can help facilities reduce consumption and hit their emission targets in order to avoid fines. Third, sustainability is often cost effective, especially in terms of energy conservation. Although sustainable practices often involve significant initial financial investments, such as the installation of solar panels or incorporation of digital information systems intended to reduce the need for printouts, they can result in dramatic long-term savings in electricity, heating, water, and office supplies. There are other sustainable practices that don't require a meaningful financial investment, including using shades and blinds, closing fume hoods, turning off lights in unused rooms, and unplugging equipment when feasible.

Hierarchy of Conservation

Hierarchy of conservation is a system used to increase positive effects on nature by taking steps to ensure resources will not be depleted or wasted by negative effects. The purpose is to establish a measurable desired goal (of a positive effect on nature) and use the following steps to achieve it: (1) avoid; (2) minimize; (3) remediate; and (4) offset.

Avoid refers to attempting to avoid harm when possible. Not all harm may be removed, but some methods may be eliminated or modified to achieve a lower negative impact. Minimize refers to attempting to minimize the current negative impacts of a plan by alternating current practices and making adjustments to methods and means. Remediate refers to remediating negative impact from past processes. Establishing and practicing rules and procedures that address areas of negative impact are effective remedial practices. Offset refers to offsetting any remaining negative impacts.

Reuse

Reuse refers to reusing items that may no longer be used for their original intended purposes. The purpose is to find another use for the item instead of discarding it. Finding discarded furniture and restoring it so that it may continue to be used as furniture or for some other purpose is a good example of reusing an item. This practice is a great way to manage damage to the environment by reducing waste and supplier demand for a product (the reused item may meet that demand). Depending on the material of the items, decomposition time can range from anywhere from 2 months to over 500 years. Practicing reusing techniques allows for a reduction of the negative impact on the environment and may produce a positive impact concurrently.

Recycle

Recycling is the act of repurposing trash materials that would have otherwise been discarded into a landfill. Discarded materials like paper, cardboard, metal, glass, wood, and plastic can often be used to make new items rather than extracting and utilizing raw materials to do so. This practice reduces waste and creates space in landfills, conserves materials like wood that are slow to renew, reduces methane production (therefore decreasing greenhouse gas production), and reduces energy production. Recyclable products can be sold domestically and globally, creating jobs and goods. Many residential and commercial buildings have municipal or county recycling services that will collect recycled items, similar to garbage collection. A number of chain grocery stores will take back plastic grocery bags for reuse. Cities, universities, airports, and other large domains are also making the practice of recycling more accessible by setting up single stream recycling containers in public locations, similar to garbage containers.

Single stream recycling allows any recyclable material to be placed into the same container for collection. In previous years, consumers often had to separate recyclable goods themselves by material, which limited the number of people who chose to recycle. Products made from recycled materials are often noted as such. For example, cardboard food boxes, fast food napkins, paper towels, and soda cans often display information that they are made from previously recycled products, and from where the product was recycled (a program, a facility, or a consumer).

Reduce

Reduce refers to reducing the items that are being used and reducing the waste that is produced. If there are fewer items being used, then there is less demand for production of items (and less waste in the supply/demand chain). If there is less waste being produced, then there is less waste to reuse or recycle. Reduction may be completed in three steps: (1) using multipurpose items; (2) using durable disposable items; and (3) limiting use to only necessary consumption and usage.

Using multipurpose items allows for items to be used for more than one purpose and reduces the need for additional items (and reduces supply/demand). Using durable disposable items refers to using disposable items that may be used more than once and for longer periods of time. Necessary consumption and usage refer to using items that are needed and not just convenient; the item must be proportional to the needs of the purpose.

Environmental Management System Standards

Environmental management systems standards have been established by the International Organization for Standardization (ISO). This organization has developed and published various industrial and commercial standards for organizations. Under ISO 14000, there is a framework of policies and procedures to assist companies in managing environmental damage that it may cause and addressing matters in an integrated manner. Companies that engage in this framework may be certified regarding ISO 14001 practices. The ISO 14000 Family refers to the standards that are developed under its framework.

Under 14001, the framework is established, and companies may follow its standards to manage environmental effects. The considerations include, but are not limited to, waste management, water and sewage matters, and air pollution. The framework is flexible and may be applied to various organizations to provide effective environmental management to the organizations and their relevant interest holders, such as shareholders. Other subsequent Family standards target specific subject matters, such as audits, communications, and specific environmental challenges that an organization may face. Additionally, ISO 14001 may be integrated into other ISO management systems, such as 45001 and 19011.

Under 45001 standards, organizations focus on workplace safety and relevant risk reduction. This allows the organizations to establish safer working conditions for employees. These standards are like OHSAS's 18001, but risk management is more prioritized than hazard management. The 45001 standards evaluate internal company matters and external considerations that are relevant to interest holders. This evaluation also includes the considerations and environments related to an organization's consumers. The internal and external evaluations allow organizations to gain an objective perception of their functions and see how each one correlates with the others and with the conditions presented.

Under 19011 standards, organizations may audit their management systems. Audits may be conducted internally or externally, and these standards provide practical procedures for conducting audits. Practical procedures include, but are not limited to, auditing principles, systems audits, audit (and employee or team) competency, and audit program evaluations. Finally, these standards help provide organizations with tools and information to assist in effective auditing practices to ensure desired performance.

Waste Removal, Treatment, and Disposal

Chemical Hazardous Waste

The following are basic requirements to follow for the disposal of chemical hazardous waste:

- Select an area near the source of the waste, out of the way of normal lab activities, and easily accessible to all lab personnel.
- Label the area "Danger—Hazardous Waste" with the following sign:

- Make sure that containers into which the chemicals will be discarded are stable enough to hold them—chemicals must not weaken or dissolve the material of the container.
- Acids and bases cannot be stored with metal
- Hydrofluoric acid cannot be stored with glass.
- Solvents (i.e. gasoline) cannot be stored in polyethylene containers, such as a milk jug.
- Waste containers must come with lids and caps that are resistant to leakage, and containers should be closed at all times, except when opened to add more waste.
- The size of the container should be appropriate for the amount of expected waste.
- Waste containers should be placed inside a larger, empty container to catch any waste that may potentially spill or leak.

Complete and attach the following hazardous waste tag to all containers:

Hazardous Waste Tag

Front

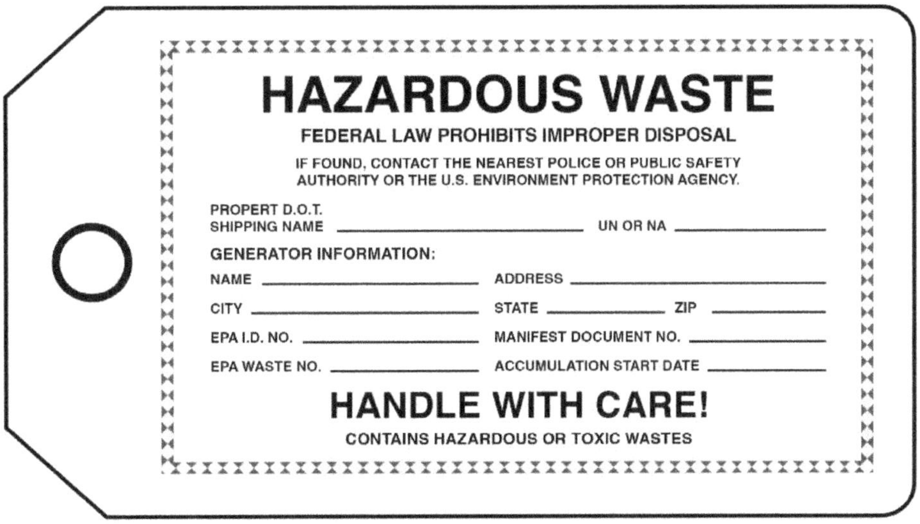

HAZARDOUS WASTE

FEDERAL LAW PROHIBITS IMPROPER DISPOSAL

IF FOUND, CONTACT THE NEAREST POLICE OR PUBLIC SAFETY AUTHORITY OR THE U.S. ENVIRONMENT PROTECTION AGENCY.

PROPERT D.O.T. SHIPPING NAME ______________ UN OR NA ______________

GENERATOR INFORMATION:

NAME ______________ ADDRESS ______________

CITY ______________ STATE ______________ ZIP ______________

EPA I.D. NO. ______________ MANIFEST DOCUMENT NO. ______________

EPA WASTE NO. ______________ ACCUMULATION START DATE ______________

HANDLE WITH CARE!

CONTAINS HAZARDOUS OR TOXIC WASTES

Back

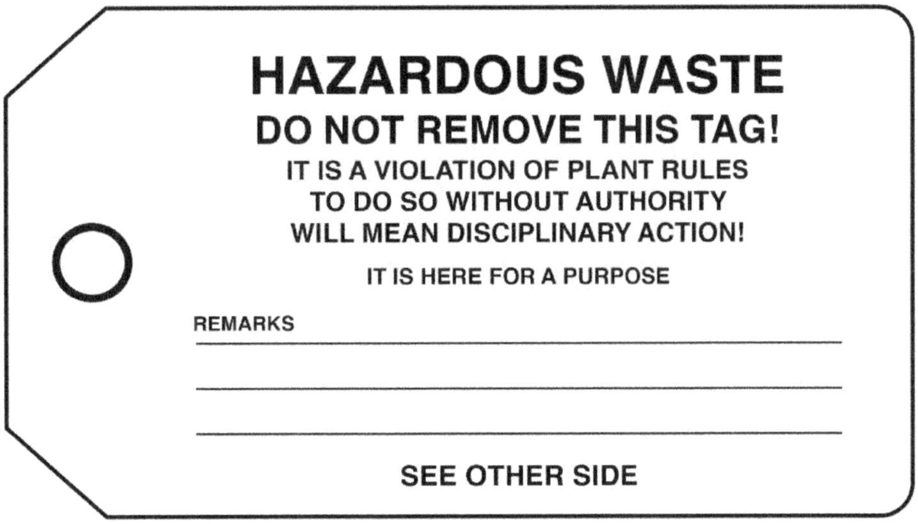

HAZARDOUS WASTE

DO NOT REMOVE THIS TAG!

IT IS A VIOLATION OF PLANT RULES
TO DO SO WITHOUT AUTHORITY
WILL MEAN DISCIPLINARY ACTION!

IT IS HERE FOR A PURPOSE

REMARKS ______________

SEE OTHER SIDE

Requirements for Liquid Waste

- Don't overfill containers—be sure to leave at least 10% space between the container opening and the surface of the waste.
- Never mix liquid and solid waste.
- Double-bag small containers, such as vials, in clear plastic bags.
- Bag small containers composed of the same kind of waste.
- Attach a completed hazardous waste tag to all bags and containers.

Requirements for Solid Waste

Chemical solid waste is composed of three different categories: lab trash, dry chemicals, and sharps.

- Lab trash
- Use for waste such as Kim Wipes, disposable gloves, paper towels, and wooden stirrers
- Double-bag in clear bags
- Attach a completed hazardous waste tag to all bags
- Dry chemicals
- Return the chemical waste to the original container in which it was purchased.
- Attach a completed hazardous waste tag.
- Sharps
    - Examples include glass (broken or intact), pipettes and pipette tips, needles, X-ACTO™ knives, or anything capable of piercing, cutting, slicing, or tearing human flesh.
- Discard any used sharps into a designated sharps container with a biohazard sign.

Remediation

Remediation refers to the removal of contaminants in the environment. Most commonly, remediation addresses pollution runoff and contaminants in soil, sediment, and water systems. The Environmental Protection Agency (EPA) regulates remediation standards, which are collectively published as the Preliminary Remediation Goals (PRG). Furthermore, state and local authorities often promulgate additional rules and procedures for remediation in order to protect the environment, ecosystems, human populations, and/or other living organisms.

The first step in remediation is to conduct a site assessment, typically in the form of a Phase I Environmental Site Assessment. This type of assessment considers the present conditions as well as the history of the site, including all the associated materials and/or production. Based on the assessment, safety professionals conduct sampling and chemical analysis to more specifically determine the type and degree of contamination.

Following the assessments and tests, safety professionals select the appropriate techniques and technologies for remediation. Remediation techniques can be broadly categorized as in-situ and ex-situ. **In-situ remediation** seeks to treat the contaminants without significant removal, while ex-situ remediation involves the extraction of contaminants and treatment of the surrounding surface area.

One of the most common in-situ remediation techniques used to treat contaminated soil is called solidification and stabilization, and it involves adding reagents to increase chemical stability and contain the contaminants within a solid area. Another common type of in-situ remediation used in both soil and groundwater contamination is nanoremediation and bioremediation. These techniques involve using nano-sized agents or microorganisms to degrade contaminants. The most common methods of **ex-situ remediation** are excavating soil and dredging bodies of water. In both instances, the soil is removed, deposited in a regulated landfill, and then treated, for example with chemical oxidation. In addition, the pump-and-treat technique is used to reduce the contamination of water. Once the water is pumped out, it is channeled through pipes containing filtering and purifying materials, such as sand, activated carbon, and chemical reagents.

Training, Education, and Communication

Training

Worksite training is integral to worker safety, and the ASP will use the available resources to ensure the worksite safety program is functional and consistently evolving. Worksite presence and intuitive interaction with workers by the ASP creates a dialogue with workers, while demonstrated empathy secures a connection. This connection manifests an increasing bond between worksite personnel. Once this bond is established, workers begin to stray from their comfort zones, which encourages honest dialogue and interpersonal conflict resolution, where individuals with disparate ideas begin to find common ground with one another. The desired end product is for people to connect on a positive emotional level.

In classroom settings, the ASP may be qualified to carry out certain training. If consultation or additional teaching resources are needed, the employer is obligated to provide them. The ASP should strive to further their certifications and training qualifications, as in-house trainings performed by an individual intimately aware of the worksite is preferred. It is equally important for the ASP to know the jobs that workers are performing in order to engage workers intelligently. Online resources are numerous, and the ASP will know and be able to reference these resources for training purposes. Assisting workers not versed in or overly familiar with online applications is the duty of the ASP.

OSHA and approved trainers provide online courses and training materials. On-site training includes videos, demonstrations, work observation, drills, and meetings. Off-site training may include classroom trainings, on- or off-site demonstrations, certification courses, and consultations.

No matter the diversity of languages, cultures, and educational levels on the worksite, employers must establish and maintain the programs necessary to achieve at least minimum-approved levels of operational safety that account for every worker on-site. In-house training may include worksite demonstrations, cross audits, presentations, drills, or safety meetings. Unique training needs may call for outside resources including consultants, specialists, and classroom courses.

Written Communication

Communication is the means by which all people receive and distribute information. Communication styles can be both verbal and nonverbal in nature. Most organizations rely on many forms of communication to accomplish their goals and objectives. Information distribution is accomplished through a vast array of communication strategies.

Safety professionals are involved in a myriad of writing activities. The goal for each is to convey information in a manner as effective as possible for the prospective audience. The best communication strategy can vary depending on the situation. **Written communication** (e.g., emailing and texting) is often chosen for convenience and the speed at which messages can be sent to recipients. A notable advantage of written communication is that the content is more clearly defined than most verbal/oral conversations. With oral and signage strategies, elements can lack detail and be missed or forgotten since there is no opportunity to revisit the material. Oral communications can require additional time and coordination to disseminate messages to all parties who need to be informed. Following is a list of the mode and purpose of written information distribution methods:

- **Memos:** Written documents that include a clearly defined purpose and a summary stating the main intention. The body of discussion conveys the writer's message and may provide a short background with facts for the reader. Latter portions of the document should state a desired outcome with suggestion(s) regarding follow-up and recommendations. Memos should be one to two pages in length.

- **Lab and Field Reports:** Written documents designed to convince the reader of the validity of results and findings. A report's outline contains a purpose, introduction, and background section providing factual data. Subsequent sections include a description of steps and methods used. The report concludes with results and findings (e.g., graphs, charts, comparative analysis, etc.) and proposed recommendations. For the reader's convenience, an executive summary containing a condensed form of the major report aspects should precede the main report. An **executive summary** can be used when communicating large amounts of information to high-level associates. This is a brief summarization of important aspects from a larger document.

- **Incident/Accident Investigations:** Written documents to determine the primary or root cause of an incident and how to prevent recurrences. Additional purposes include fulfilling legal requirements, workers' compensation processing, assessing levels of compliance, and determining associated costs.

- **Meeting Minutes:** Modifiable documents that cover meeting agenda items and include names of participants and attendees, responsible personnel, notes pertaining to subject matter, date and time duration of meeting discussions, and action items discussed or addressed.

- **JSAs:** (See previous section covering JSAs.)

- **Surveys and Questionnaires:** A series of calculated questions (designed for yes/no answers or responses) applied to a rating or scoring system. Subsequent steps include data compilation and analysis to determine program and cultural successes and failures. Includes proposed recommendations to address deficiencies.

- **Operating Procedures and Instructions:** Written documents that provide step-by-step directions to learn an operation, complete a task, or perform an assembly. They consist of a combination of writing, pictures, labels, and symbols.

- **Data and Trend Analysis:** Gathering and analysis of data related to past events and behaviors over time for the prediction of future outcomes. This is achieved through tracking deviations, variances, and repetitious elements.

- **Emails:** Most notably, performance with writing emails is a matter of etiquette. Writers should keep subject matter to the interests of work and avoid criticizing or responding when angry. Writers should also be brief, formal, and proofread for errors and required edits. The writer should include a subject line appropriate to the content of the email.

Signs and Symbols

Safety in construction activities incorporates a vast array of **signs and symbols**. These methods of communication are used to relay important cautionary information to both involved and uninvolved personnel. Hazard Communication Standards, lockout/tagout, crane and heavy equipment operations, and electrical safety are among the many processes where the use and application of these strategies are of vital importance.

- **Labels and Identifiers:** These elements are used to communicate elements such as ownership, manufacturer's criteria, expiration dates, potential hazards, or weight and load capacity.
- **Pictograms:** These are used as a visual communication method to caution or warn about various conditions. Pictograms are advantageous because of their capacity to defeat language barriers and achieve universal application.
- **Hand and Arm Signals:** These are physical movements/motions performed to provide assistance and guidance to help perform safe actions. Hand and arm signals are used to give commands, warn of impending hazards, increase alertness, and eliminate the interference of multiple processes.
- **Color:** Color is used to differentiate identity and function of otherwise similar components. Examples include the use of black, white, and green for wiring, and blue versus yellow to differentiate a water line from a gas line.
- **Alarms and Barriers:** These are used to warn, gain attention, and/or provide separation or isolation from an existing hazard. They employ device elements that impact multiple senses such as: sound and light amplitudes and frequencies; color; and placement with respect to lights and alarms, flags, lines, and cones.

Adult Learning Theory and Techniques

Adult learning theory is known as andragogy, and there are significant differences between andragogy and childhood learning theory (pedagogy). Compared to children, adults have a significantly broader and deeper base of pre-existing knowledge, which is both a benefit and a detriment. On the one hand, the knowledge serves as a strong foundation for acquiring new concepts and skills, but on the other hand, adults might be relying on inaccurate or incomplete information. Additionally, adults have significantly more experience with self-evaluation, and they are more likely to be directed by internal motivations. Therefore, andragogy is largely based on the principle of self-direction. Andragogy also emphasizes learning through problem-solving and practical application to underscore the importance and usefulness of what is being taught, which increases adults' motivation to learn. Lastly, andragogy treats adults as respected partners in the learning process. Adult education is individualized, and it focuses on providing guidance about how to self-evaluate and when to take different approaches.

The most effective adult learning techniques usually incorporate the principles of andragogy. Broadly speaking, a useful technique is to present the benefits and practical importance of the learning objective before the planned lesson itself. Once adults recognize the benefits of the lesson, they will be significantly more motivated to achieve the objective. Adults usually have a grasp on whether they are visual, auditory, or kinesthetic learners; so adult education programs should individualize the learning process whenever feasible to increase its efficacy and maximize retention. Furthermore, self-directed learning objectives serve as a vital learning technique in adult education.

When using this technique, instructors provide learners with an objective and suggested resources, such as websites or tutorials. Self-directed learning respects the independence of adults by removing unnecessary steps and allowing them to work at their own pace. Additionally, self-directed learning exercises can increase workers' participation, enthusiasm, and motivation by actively involving the worker in the learning process, particularly in terms of providing the worker with more freedom in how to pursue and achieve the learning objective. Although self-directed learning heavily emphasizes independence, safety professionals must still play a supervisory role to ensure the work is remaining on task and progressing toward the goal.

Another effective technique is experiential learning, which means learning by doing. In most instances, experiential learning involves providing workers with opportunities to actively solve problems in an authentic and meaningful context, like interactive demonstrations, simulation, or practice attempts. Most experiential learning techniques consist of three stages. First, the worker acquires concrete experience with the subject matter, typically through hands-on learning. Second, the worker reflects on the experience, either independently or with coworkers. For example, the worker might write a summary, draw a comparison with prior experiences, and/or discuss the experience in a small group. Third, the worker applies what they've learned in a real-world scenario, such as a problem-solving exercise, case study, or simulation. Experiential learning is most often used during on-the-job training, meaning the learning process occurs within the working environment. An experienced worker or manager usually implements on-the-job training by teaching the worker about processes and how to use the machinery and equipment. On-the-job training is generally considered to be a highly effective learning technique because experiential learning involves the exact type of processes and tools the worker will be using every day.

Presentation Tools

Oral communication is used to verbally transmit messages to an individual or group. Oral communications can be either formal or informal. Traditional types include face-to-face conversations or meetings, speeches, and telephone communications. Technology continues to offer new ways to transmit oral communications. Video communications such as Skype™ and podcasts are becoming more common in the workplace.

The most effective form of oral communication is considered face-to-face (in person) communication because it reduces the number of obstacles (also referred to as "noise") that impede a recipient's understanding of the message. This form allows for timely two-way communication versus the undesirable component of a delayed exchange in one-way written content. Face-to-face communication also allows for immediate feedback. Its flexibility in reiteration can result in improved understanding when compared to written communications.

With face-to-face communication, the deliverer of the message has the opportunity to assess the audience by observing reactions and has the ability to immediately confirm that the audience understands the message being delivered. In turn, the audience is able to ask questions and receive immediate feedback from the deliverer to clarify the message. By contrast, written communication does not afford the communicator the same opportunity to immediately reiterate or clarify the message being communicated. Situations usually exist where recipients may not be able to solicit the additional written information necessary for understanding the message. The following list includes the mode and purpose of oral information distribution methods:

- **Meetings:** Scheduled events to discuss and exchange information on a specific topic in a group setting. Meetings can be formal or informal.
- **Toolbox Talks:** Pre-task discussions on specific safety issues or the review of information before performing an altered task or undertaking a new process. Toolbox talks focus on one topic per discussion.
- **Three-Way Communications:** A process safety communication method where an actuating member verbally states their command of action to warn others. Other member(s) verbally respond in a similar way to confirm recognition. The actuator repeats the command before performing the action.
- **Presentations:** Provide training, guidance, or familiarity to a group on a particular topic.
- **Speeches:** Convey/Broadcast message(s) to a larger group.
- **Telephone Conversations:** Used to promptly provide information to an individual or group.
- **Interviews:** Performed to investigate or assess a person or incident.
- **Video Conference:** Used to share and exchange information with remote participants so they can see and hear other participants.

Safety Culture/Climate

The foundation of effective safety is based on an in-place culture of organizational safety. This can be described as all levels of an organization holding common beliefs for the prevention of injury and illness. Organizations have the duty of providing a safe and healthy workplace to employees or, as stated by the General Duty Clause, a workplace that is "free from recognized hazards causing or likely to cause death or serious physical harm." This mechanism ensures the input of critical elements for site safety plans, processes, and performance.

Health and Safety Programs

To protect the health and safety of employees, core elements make up the management philosophy and are the strength of programs in place. Since success starts at the top of an organization, effective safety and health programs should reflect this philosophy as well. Safety and health programs should be led and supported by management's commitment to leadership and accountability. Management develops written safety plans, sets organizational goals, and should set a good example for employees. An effective safety and health program should consist, at a minimum, of the following primary elements:

- Management commitment
- Document control
- Employee involvement
- Hazard detection, assessment, and prevention with a hierarchy of controls
- Support relevant to contractors and subcontractors
- Worksite assessment and policy audits

- Continuous improvement with tracking and performance measures
- Daily and periodic inspection of machines and equipment
- Communication and training with accountability and enforcement

The application of these elements results in programs with clearly defined goals and reflects management philosophies with organizational accountability from top to bottom. Employee involvement and feedback is encouraged and helps to promote a positive safety culture. An "all-in" attitude serves to integrate safety and health goals into the organization's structure. Long-term tracking and solutions support and ratify continuous improvement efforts. Through collaborative efforts and imagination, cost-effective solutions can be presented that support an organization's bottom line.

Safety programs should consider health and medical aspects to meet the needs of the workforce. These include elements of nursing, first aid, disease control and prevention, and industrial hygiene. Health and safety programs should do the following to be successful:

- Maintain a healthy work environment
- Support disease diagnosis and immunizations
- Maintain an effective workers' compensation program that supports adequate healthcare and rehabilitation for work-related injury and illness
- Support open communication between employees' doctors and organizational health personnel
- Ensure periodic baseline testing and monitoring with respect to environmental exposures
- Provide drug and alcohol testing to protect personnel
- Provide health and fitness education and counseling
- Tailor health program elements to work processes
- Ensure workers' physical and mental capabilities are acceptable and fit for their job

Types of Safety Programs

An employer must make an honest and comprehensive assessment determining the individual mindset and the collective pulse of all project personnel. Behavior-based safety programs (BBSPs) will be key in the mitigation of assessed challenges and may include:

- A BBS that discerns the association between human behavior and injury prevention.
- A perception of how the attitudes and emotional state of workers influence their behavior in the workplace. It is the purpose of the BBSP in place to facilitate an avenue for constant exchange of official and informal information inclusive to all personnel.

Motivation is the incentive that drives people to behave in certain ways. To motivate employees to work toward the goals of the organization, leaders must understand what motivates people.

Motivational theories seek to explain what motivates people so that organizations can direct workers to behave in desired ways. Motivational theories can be content oriented or process oriented.

Content Oriented

Content-oriented theories seek to explain why people behave in certain ways.

Maslow's Hierarchy of Needs: As people fulfill their basic needs, they strive to fulfill greater needs higher up the hierarchy.

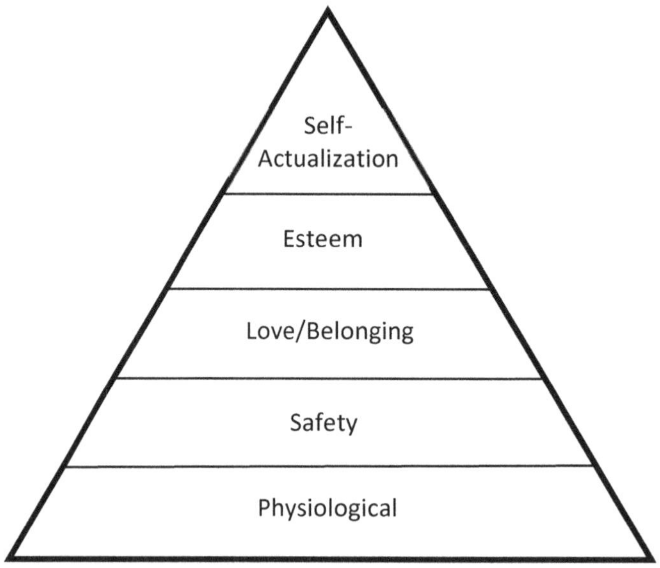

McClelland's Model of Realization: People are motivated by the need for relationships, power over their surroundings, or the ability to achieve goals.

Process Oriented

Process-oriented theories seek to understand how processes and psychological factors affect motivation.

Vroom Expectancy Theory: People are motivated by the value of the reward they will receive from achieving goals. These rewards can be internal (fulfillment) or external (promotion).

Goal-setting Theory: People are motivated when they set specific goals that are challenging to them and encouraged by the organization. Motivation is sustained through the feedback they receive as they work to achieve their goals.

Data Collection, Needs Analysis, Gap Analysis, and Feedback

Data Collection

Data collection methods will vary based on the type of solution or process that is needed. Surveys can be used to gather individual feedback from a wide demographic. They can be conducted in person, online, or over the telephone; however, they are subject to interviewer bias, voluntary completion, and low completion rates. Focus groups utilize the services of a skilled facilitator who solicits feedback and opinions about a specific topic from identified stakeholders. Focus groups run the risk of low engagement or facilitator bias. Observational data collection utilizes one data collector to observe a specific context and take notes; however, this is highly subject to bias if the data collector is visible. Other types of data collection can be quantitative. If an organization maintains diligent data management practices, these can usually serve as easily accessible sources of information. Permission to use personal information is normally required.

Determining Training Requirements

To determine training requirements for work site personnel with varying levels of skill, education, experience, and language proficiency, leaders conduct the following assessments:

- Organizational
- Individual
- Training Needs

The organizational assessment examines the organization to determine the current level of performance of the organization to identify areas in which performance can be improved. The assessment also measures which skills, education, and knowledge are required in order for the organization to achieve the desired level of performance. This assessment helps identify strengths and weaknesses in the organization, so leaders can enhance their strengths and work to eliminate weaknesses.

The individual assessment examines the performance of the workforce. Leaders assess groups or individual workers in the areas in which performance requires the most improvement. The assessment identifies which groups or individuals require the most resources and which specific skills they are lacking. During this assessment, leaders may determine all the workers in a particular group are lacking specific skills, or that only one or two require additional training.

The organizational and individual assessments identify the specific skills and knowledge the workforce requires to perform at the desired level, and which groups or individuals require the most training. The training needs analysis determines how to fill the gaps in knowledge.

Once the training needs have been identified, leaders must identify the best way to deliver training based on the limitations of the workforce. Workers who lack prerequisite skills might require additional training before they can train with the rest of the workforce. In this case, leaders might determine that the workers can gain the knowledge they need from computer-based training or a one-day seminar. Some workers might require training in different languages, or they might be spread out in different geolocations. Leaders might decide to create computer-based training courses recorded in multiple languages that are accessible anywhere in the world.

Needs Analysis

A **needs analysis** is the process in which an organization gathers information about the principal needs and requests of its members. This analysis studies the expectations and requirements of subjects who are affected by workplace programs or regulations. Such individuals may include employers, teachers, administrators, donors, and family members of students. Needs analysis results may be used to clarify the objectives of an organization, or as a teaching tool in a classroom.

A needs analysis typically starts by gathering data. This process can be accomplished through a multitude of channels such as surveys, interviews, questionnaires, or polls. In a needs analysis, problems and inefficiencies are clearly identified. These issues are ultimately addressed by the organization through the implementation of improvements to maximize results. A needs analysis may serve as an efficient means of examining organizational procedures and techniques of training at minimal cost. Additionally, a needs analysis can be a helpful tool to develop occupational injury prevention programs within an organization.

If the needs analysis is conducted properly, the organization's next step is to implement the suggested changes in a way that promotes success. This endeavor requires the allocation of resources and personnel to the proposed plan. The proposed changes should meet the organization's productivity targets and fulfill the requirements of governmental agencies.

Gap Analysis

Gap analysis is the analysis of a company's current performance in comparison with the company's desired performance. Conducting a gap analysis will allow a business to increase its efficiency, production, and profits by determining "gaps" within the company. It is performed by: (1) determining current performance; (2) determining desired performance; (3) determining gaps within performance; and (4) determining policies and procedures to minimize or eliminate gaps.

Determining current performance involves the evaluation of data and conditions within a company's performance system. This allows for an objective review of the performance system that shows where the company is succeeding and where it needs to improve.

Determining desired performance is an identification of how the company intends to perform either currently or in the future. Determinations should be fact specific and measurable to ensure that the desired performance is being obtained.

Determining gaps within performance shows the difference between current performance and desired performance. This identifies conditions that hinder desired performance.

Determining policies and procedures to minimize or eliminate gaps between current performance and desired performance allows for a company to actively pursue desired performance and *bridge the gap.*

Interpersonal Communication and Feedback

Interpersonal communication is one of the most valuable components of communication and group dynamics in the workplaces because it facilitates the development of processes, maintenance of operations, and establishment of goals. Generally speaking, interpersonal communication involves a communicator delivering a message to a targeted audience who, in turn, act as communicators to deliver feedback, which can be used to gauge comprehension and interpretation. Communicators must be sure to limit the noise, which is a distortion in the message. Common causes of noise in messages include technical jargon and vague generalizations. When noise obscures the message, workers are significantly less likely to be properly aligned with the message and effectively perform job functions.

One of the best ways to enhance interpersonal communication is to improve listening skills. Strong listeners tend to have a positive attitude and an open mind, use body language to demonstrate they're paying attention, maintain active engagement, acknowledge the message, provide appropriate feedback, and synthesize content with prior knowledge and experiences.

Feedback is a quality tool in the communication process between a company and its employees. Companies and its personnel convey messages to employees through various forms of communication, both verbal and written. The speaker is the individual giving the message, and the listener is the recipient of the message. The speaker may believe that the message is clear and covers all matters that need to be addressed. However, the listener may not understand a message to its full intended extent.

Feedback allows the opportunity for a listener to provide additional insight on the message and ask questions to better understand the message. Responding to feedback allows the opportunity for the speaker to clarify the intended message and increase the listener's comprehension of the message. This process allows for a clear message and the elimination or reduction of communication barriers that may affect the message. Providing unambiguous communication allows for the speaker and the listener to effectively comprehend the message in a similar manner and can lead to increased performance.

Assessing Competency

Competency assessments evaluate whether someone has the knowledge and skills necessary to effectively and safely complete a specific task. Often, competency assessments are legally required, especially when workers are dealing with hazardous substances and dangerous equipment. In addition, competency assessments play an important role in reducing workplace accidents, and some insurers require assessments or certification to maintain policies.

Worksites most commonly conduct competency assessments during training either after the completion of a preliminary stage and/or the entire program. Other times training assessment occurs after a prolonged period during which the worker hasn't actively applied the relevant skills and knowledge, especially if the worker has been observed operating equipment incorrectly or unsafely. Some worksites and government regulations mandate that competencies be obtained within a certain time frame. In addition, changes to worksite conditions frequently trigger competency assessments, including but not limited to the passage of new workplace regulations, replacement of equipment, alteration of tasks or processes, or the occurrence of accidents and near misses.

Types of assessments vary, but they all feature an objective standard to measure success and failure. Examples of frequently used competency assessments include written tests, verbal tests, and practical demonstrations. Adults generally prefer practical demonstrations of skills and knowledge because the demonstration is more directly related to performance in the field. However, written tests and verbal tests can serve as appropriate alternatives when a practical demonstration isn't feasible. Written and verbal questions should prioritize essential knowledge and skills rather than extraneous and theoretical information.

An important component of competency assessment is the recording of all results, which has three valuable benefits. First, the results of assessments provide quantitative information about how many workers can accomplish tasks, which increases the efficiency of production. For example, competency data can help optimize scheduling and prevent stoppages in workflow. Second, assessment data can be used to identify workers who will need corrective action to achieve competency. For example, if a worker repeatedly fails an assessment, the underlying issue must be addressed, including adjustments to the training program. Third, when taken as a whole, assessment results help evaluate the efficacy of training and education problems. If only a select few of employees fail to achieve competency, then the issue is likely best addressed individually. However, if a majority of workers aren't acquiring the requisite skills and knowledge in a timely manner, the program itself might be flawed. As a result, assessment results serve as critical data for adjusting and updating training programs. In order to reap these benefits, safety professionals must have a centralized data management system for collecting, compiling, and analyzing the assessments.

Legal Liability

Legal liability is the obligation imposed on a person, or artificial person (business), that is held accountable for causing harm to another person, or artificial person. The imposed obligation is generally expressed in monetary value. There are two overarching sources of law regarding legal liability: (1) criminal liability and (2) civil liability. **Criminal liability** is liability imposed by a government proceeding to redress an applicable legal violation. **Civil liability** is an obligation imposed on a person, or artificial person, to either pay monetary compensation or perform a specific conduct imposed by the result of a lawsuit. Civil violation may arise from a contract or tort claim.

Contract is a legal relationship between two or more people that agree to do or not to do something. The failure to follow the terms of the agreement may lead to a breach of contract. Breaches may occur at various times within a contract, but generally result during the performance or discharge of an obligation stage of a contract.

Tort is wrongful conduct that has infringed upon the rights of another person, or artificial person. These claims may be applied to employee-employer incidences. A common type of tort claim is negligence. **Negligence** is the breach of a duty that was owed to another, and that breach resulted in a harm to another. The resulting harm is the civil liability that must be addressed by the breacher of the duty. However, there are legal defenses to a negligence claim: (1) assumption of risk; (2) contributory negligence; and (3) fellow-servant rule. Assumption of risk implies that the harmed person, or artificial person, accepted the risk when in the scope of performing a specific conduct. Contributory negligence suggests that the harmed person, or artificial person, contributed to their harm when they conducted themselves in a certain manner. Fellow-servant rule means that an employer is not liable to an employee for a harm that resulted from other employee(s). A type of tort law matter is workers' compensation.

In 1911, the first workers' compensation laws were passed in Wisconsin, followed by ten other states that same year. By 1948, all states had enacted these laws. Workers' compensation laws are codified rules regulating employee-employer matters and resolutions. These are specialized with the intent to be applied specifically to employer-employee matters and to provide clear expectations and promote efficient resolutions. These laws attempt to create win/win scenarios in which an employee is compensated, and an employer may continue to operate a business.

Work related injuries may be categorized based on the resulting harm: (1) partial; (2) total; (3) temporary; and (4) permanent. Partial injury is when an injured employee is able to work but not able to perform all the duties required of the position. Total injury is when an injured employee is substantially unable to perform the required duties of the position, or none at all. Temporary injury is when an injured employee is expected to make a full recovery. Permanent injury is when an employee's injuries are irreversible, and the employee will not recover. Injuries may qualify for one or more categories.

Most jurisdictions require an employer to be self-insured or maintain workers' compensation insurance if it maintains a minimum number of employees determined by that jurisdiction.

Ethical Behavior

Professional Practice

The BCSP—along with the Council on Certification for Health, Environmental, and Safety Technologists (CCHEST) and American Board of Industrial Hygiene (ABIH)—are an alliance utilizing their collective expertise to increase the levels of education and proficiency in the workplace. These are some of the organizations that, in concert with OSHA, promote workplace professional development and pursuit of approved certifications. The ASP is encouraged to participate in the safety community and engage in higher learning opportunities, certification options, and the professional network the community provides.

The Board of Certified Safety Professionals (BCSP) is a peer certification board that sets standards for safety, health, and environmental professionals. Its core values are respect, accountability, leadership, and excellence.

The BCSP evaluates the technical competency, experience, performance, and educational credentials of professionals. It also administers exams to certify that professionals meet the standards defined in the certifications, and it continuously monitors performance to ensure certified professionals uphold those standards in the course of their work and their interactions with colleagues, clients, authorities, and the public.

The BCSP grants the following certifications:

Construction Health and Safety Technician (CHST)
Certified Environmental, Safety, and Health Trainer (CET)
Certified Safety Professional (CSP)
Associate Safety Professional (ASP)
Occupational Hygiene and Safety Technician (OHST)
Safety Trained Supervisor (STS)
Safety Trained Supervisor Construction (STSC)
Graduate Safety Practitioner (GSP)
Safety Management Specialist (SMS)

A certified ASP is subject to the BCSP's Code of Ethics and professional standards.

Audits

Audits of health and safety programs promote the health and safety of a site's employees, vendors, and guests. The audit allows for identification and assessment of hazards and development of procedures to eliminate, or at least reduce, hazards. It also requires determining whether such procedures are effective towards achieving a desired outcome. Audits may be conducted either by: (1) document review; (2) employee interviews; or (3) site condition evaluation. Records may show the determination and frequency of issues that may occur on a site. Additionally, this may be corroborated by employee interviews. Finally, objective site evaluations (seeing a site in person) assist in building a better comprehension of current and potential issues. A walk through of a site is one of the best resources in continuously evaluating the health and safety of its employees, vendors, and guests. Managerial commitment, employee involvement, worksite analysis (analyzing site conditions to address current and potential hazards), hazard policies and procedures, and training are imperative to achieve audit goals.

Record Keeping

Documenting safety information and processes is a critical part of safety. OSHA, other governing agencies, as well as insurance providers, require documentation. Documentation is essential for control, tracking, and maintenance of various safety processes and will be requested in the event of an external (OSHA) post-incident inspection or safety audit. A multitude of safety processes and supporting elements are required by law to be written and documented. Such items include confined-space procedures, emergency action plans, respiratory protection, and hazard risk assessment. OSHA states that organizations should consider the following elements for achieving regulatory compliance:

- Knowledge of standards applicable to an organization's work site and program needs
- Documentation for certification of hazard assessment including the particular site or facility, hazards, controls, procedures, and training
- Continual maintenance, review, and updates for training records
- Completion and recording of OSHA-mandated certifications and assessments

For developing documentation, an organization should first prepare a safety mission statement that demonstrates commitment to protecting the health of their employees. Effective document development relies on teamwork from a safety committee comprising key departmental personnel that direct the collection and sensible organization of required document elements. This committee should also develop approaches for meeting/exceeding regulatory compliance and for reaching the goals and objectives of company policies.

Part of document maintenance and control includes tracking and identifying recent revisions to the document. Organizations must have a system for identifying which written plans and documents are most current. A document coordinator should be selected to control and maintain documentation. Using outdated documents results in lost time, wasted efforts, and potentially harmful impacts. Organizations should maintain documents in a centralized and secure location (e.g., electronic repository) and make them available to employees and associates as necessary.

Employers and contractors must track and measure safety performance. A critical function of program management and continual improvement is identifying trends and indicators relative to safety warnings issued, loss, injury, and near-miss incidents.

Recordkeeping includes the documentation and filing of the training that was delivered, the inspections and audits that were conducted, the environmental hazards that were monitored, the drug screenings that were performed, etc.

For application of its safety intervention strategies in the worksite, OSHA uses the injury and illness recordkeeping standard and written programs that must be produced by the employer should OSHA request them. The records the employer maintains are work-related fatalities, injuries, and illness, which include and identify hazardous industries, employers with poor safety programs, and trends for injury and illness. The ASP must realize how crucial proper recordkeeping is and how it validates a successful safety program. It also must be available for inspection during audits and is an excellent record of past hazards and their mitigation as applied to future activities. CAPA records can be included in this practice.

While small businesses (less than ten workers) remain free of routine illness and injury reporting, an increasing variety of commercial enterprises now fall under the routine and compulsory reporting for illness and injury required by OSHA. Previously discussed variances in worker ability to retain information must be identified, assessed, and remedied with the best possible means.

Sampling

Sampling refers to acquiring samples of a specific population for the use of applying it to an analysis of a subject matter or specific content. An analysis should provide an objective perception, and it is important to account for different variables that may affect the subject matter or content that is being analyzed. Sampling may apply to such things as surveys or consolidated reports.

Regarding sampling and ethics, the sampling pool should be reasonably broad enough to cover all areas that affect the content being produced. The samples should not solely be from a preferred source that does not necessarily align with ethical guidelines. Producing a sample based on preferential sampling would create a misrepresentation of the matter that is being analyzed. Ethical sampling requires taking samples from reasonable and diverse sources and compiling the information into a final product. This final sample or result may not give the desired outcome or showing, but it is honest and provides an objective picture of the subject matter and the sampling methods that were used.

Standard Writing

Standard writing of a certified ASP should be ethical. This requires that the professional should write with honesty, integrity, and be unbiased about written safety matters. The professional should not depict facts in a manner that is misleading or disproportionate to its facts. This includes past and current events, whether related to the immediate matter. It also includes facts related to employees, vendors, and guests that may be involved in the matter. The ASP should avoid the use of bias, and make sure that all standards are written to reflect inclusivity. Standard writing practices should be in manner that enhances the safety profession and provides quality information internally and to the public.

Plagiarism is the act of a person representing written work as their own work when it was actually produced by another person. **Copyright** is a legal right for one entity to use and share specified written content, with limitations on others using the content without permission. Legal processes are used to secure a copyright on registered content. A copyright is exclusive to the registering entity. Professional standard writing should not plagiarize or violate a copyright.

BCSP Code of Ethics

The BCSP Code of Ethics ensures the professionals they certify adhere to a strict code of conduct to maintain the integrity and honor of the profession.

The Code of Ethics includes the following ideals:

- Hold paramount the safety and health of the people, the environment, and the property by informing employers, property owners, clients, the public, and the appropriate authorities in a timely manner if they discover dangers or unacceptable risks to safety and health.

- Be honest, fair, and impartial by balancing the interests of the client, property owner, employees, and the public and avoiding any practices that could deceive workers or the public.

- Issue public statements objectively and truthfully, and only when the information being presented is based upon strong evidence from competent subject matter experts.
- Accept assignments only when qualified to do so by experience and education in the specific field.
- Present only truthful and accurate information about previous experiences, education, and professional accomplishments to avoid misrepresenting or being deceptive about professional qualifications.
- Conduct relationships with transparency and professional integrity to avoid conflicts of interest that could lead to compromised judgement, and advise the BCSP of any misconduct by certified professionals.
- Act without bias toward age, gender, ethnicity, religion, nationality, sexual orientation, or disability.
- Provide service in civic affairs by sharing skills and knowledge and by taking action that promotes and improves the safety, health, and well-being of the community.

Protection of Worker Privacy

In business, confidential information refers to protected information about the business that is not available to the public. Confidential information includes trade secrets, personal medical information, and personally identifiable information.

Information

Trade secrets are defined under the Uniform Trade Secrets Act as information that holds economical value for a business due to being generally unknown to other businesses or the public. This includes proprietary information such as technical specifications documents, blueprints, formulas, product designs, algorithms, programs, techniques, processes, and information such as customer or supplier lists.

Trade secrets do not include other protectable confidential information such as employee agreements, non-disclosure agreements, or procurement documents.

Personal medical information is protected under HIPAA, which states that people have a fundamental right to privacy. Patient information must remain confidential. The patient must authorize any release of their medical information, and they may withdraw this authorization. Additionally, patients have the right to access their information at any time. Only the minimum amount of information required should be disclosed during legal record audits.

Personally identifiable information is information that can be used to identify, contact, or locate individuals. This includes financial, legal, employment, and education information. This information is protected under the law in some countries and by compliance requirements in others. Compliance rules often require that this information be encrypted, protected through data loss prevention tools, and stored according to rules that dictate how much information can be stored, where, and for how long. Finally, compliance rules regulate how much of this information can be shared within the organization and with third parties.

Security for both physical files and computer recordkeeping involves the control of access and storage. A project recordkeeping system collects, classifies, and stores a significant amount of private and intercompany information for employees, clients, contractors, and any exterior personnel employed on the worksite. Personnel given access to the recordkeeping database might be asked to sign a confidentiality agreement to access such.

Dealing with Unethical Situations

If a BCSP certified professional becomes aware of professional misconduct by another BCSP certified professional, then he or she must take actions to bring that misconduct to the Board of Certified Safety Professional's attention. Professional conduct should be held to the highest standard of principles. This standard ensures that professional judgment is not hindered by potential conflicts of interest.

Conflicts of interest must be avoided especially in situations where one party would derive personal benefit from a shared interest. Safety professionals must also avoid discrimination in the workplace. Policies and procedures must be free of bias or harassment based on religion, ethnicity, national origin, gender, sexual orientation, age, disability, marital status, or veteran status. This includes any characteristic that would be protected by applicable laws within the respective jurisdiction. Professional conduct should advance the health and safety of the community and the safety profession.

Conduct should be honest and not misleading especially regarding material facts. Depending on the degree of the misconduct, remedial actions may include additional training, directly confronting the respective individual, alternate conduct suggestions, deterrent programs, or a formal complaint submitted to the Board of Certified Safety Professionals. The purpose of ethics is to discourage undesired behavior and conduct, not to punish the respective individual.

Employee Putting Others at Risk

Risk management is one of the foundations of the safety profession; therefore, safety professionals should be aware of processes that are used to address and manage risk. An employee may put themselves or others at risk. The potential harm of such actions is considered a hazard in the workplace. It is imperative that the hazard(s) be identified and examined for an effective resolution. Identification requires an immediate determination of the conduct and conditions that pose a hazard. Examination determines the likelihood and severity of harm resulting from the hazard. Reasonable policies and procedures should be developed and implemented to deter undesired conduct. This may include such things as hazard training and education requirements, safeguards and warnings, and risk avoidance. Effective policies and procedures require clear standards and guidelines as well as personnel with sufficient knowledge and skill to enforce those policies and procedures.

It is management's duty to ensure that an employee understands how to reduce or eliminate risk. However, policies should be in place that provide repercussions when an employee knowingly puts others at risk. Disciplinary policies help to deter employees from engaging in risky behavior. Consequences include, but are not limited to, employment deterrents and demerits (writeups, additional training, etc.), legal liability, or loss of income due to reduced work hours or temporary suspension.

Reading and Interpreting Regulations

Safety Legislation and Regulation

A sound knowledge of applicable standards and safety legislation and regulations (and the capability to source pertinent reference documents) for the worksite is the responsibility of the ASP.

Employers are responsible for providing a safe workplace to employees. The safety professional is responsible for maintaining knowledge of regulations and how they apply. Science and technology can impact safety methods, best practices, and other elements, which can cause current standards and codes to change. Researchers should understand that the Internet contains vast amounts of inaccurate information and nonfactual elements based on assumptions and biases. Therefore, researchers must be diligent in efforts regarding cross-referencing, fact-checking, and confirming the credibility of sources.

Enforcement includes an array of subtopics such as inspections, high-hazard alerts, penalties and enforcement for severe violators, and letters of interpretation. Training and Education includes e-tools that provide web-based interactive training to users and safety and health topics that provide information on detecting and controlling hazards. Both are designed as training tools and provide relevant OSHA standards, how the standards apply, and information compliance.

Safety and health professionals can also reference OSHA's frequently asked questions (FAQs) to gain further understanding of regulations and how they apply. OSHA FAQs address over thirty safety- and health-related topics such as inspections, falls, incident reporting, training, etc. Each topic subsection provides a wealth of information and supporting regulations helpful to researchers.

Effective communication is the primary conduit for worksite safety, and the ASP is accountable for the proper and effective dissemination of topical worksite safety, health, and environmental information. The ASP will utilize the following communication methods: industry standards publications; regulatory resources and publications; checklists; in-house and contractor Health, Safety, Security, and Environmental (HSSE) plans; HSSE signage; posted notices; and most importantly, job hazard analysis (JHA), worksite presence, and conversational engagement with all levels of personnel.

The ASP must approach this challenge with enthusiasm, impartiality, and empathy. The ASP will participate in the prioritization of worksite hazards, mentoring of worksite personnel, and corrective actions and resolution of any resultant worksite conflicts.

Relevant Current Information

Keeping up-to-date with change is vital for maintaining compliance in programs and policies. Many organizations often overlook aspects of change management and continual improvement efforts. Implementing safety elements based on standard and code modifications demonstrates commitment and due diligence. Conversely, failing to maintain and apply current knowledge results in employers with deficient programs, thereby prescribing outdated work processes, static training methods, and associated best practices. In the event of an OSHA inspection, or a program audit due to a serious incident, such practices can result in serious consequences for the employer where negligence can be a factor.

Examples of recent changes are OHSA's widely known **Hazard Communication Standard (HCS)** (1910.1200 – 1926.59) and the **final ruling on crystalline silica** exposures. Such changes impact a multitude of elements which are applicable to in-place employee programs, processes, and training methods. With respect to the silica ruling, the following are program elements subject to review and revision:

- Access control plan/regulated areas
- New engineering and administrative controls
- Respiratory protection/associated action level and permissible exposure limits
- Medical surveillance
- Hazard communication training
- Personal protective equipment

The safety professional is responsible for keeping up with OSHA standards and best practices. Hazard impacts, new developments in technology, improved methodologies, and scientific research are all part of elements that necessitate rule modifications and updates. Knowledge of the appropriate tools and resources can simplify this overwhelming process. A complete understanding of modifications, applications, exceptions, and other factors can be gained by viewing OSHA interpretations.

In today's technological environment, computers and the Internet are by far the most effective tools for accessing current information. Their main advantage over any paper publication is that updated information can be accessed at light speed. However, the sheer volume of Internet material coupled with deficiencies in searching skill can limit these advantages. Obtaining current information requires skills in the use of search terms, methodologies, and features provided within the website. Safety professionals and management can use the following sources for keeping up to date with ongoing changes in standards, codes, and best practices:

- OSHA.gov
- Safety.BLR.com
- J.J. Keller Online
- SafetyDailyAdvisor.com, available at ehsdailyadvisor.blr.com
- Other sites developed for state-run compliance

Among the tools listed above, OSHA.gov is the most complete tool a skilled safety practitioner can use to access current information. OSHA's website provides features and options including a comprehensive home page, advanced search, and a full A to Z index of topics. Data & Statistics offers searching by establishment, inspection data, and frequently cited standards by **Standard Industrial Classification (SIC)** code or industry type. Users can also obtain data related to accident investigations and fatality statistics.

Determining Appropriate Actions Based on Knowledge Limitations

A safety professional will sometimes be faced with an issue that he or she may not know how to approach. Additionally, the professional may have limited knowledge on the matter. There are three problem solving approaches that will likely help a professional with this type of issue: (1) analytical problem solving; (2) logical problem solving; and (3) rational problem solving.

Analytical problem solving allows a professional to evaluate all aspects and conditions in an objective manner to determine what the issue is. By reviewing aspects or components of the issue, the professional may foresee where their gap of knowledge may be. A sufficient fact-finding process may help to gain new insight into an issue or problem.

Logical problem solving allows for a professional to collect information and data and organize it into sequential order. The structure allows for a comprehension of the greater issue and why it may exist. Based on prior information and data, a professional may infer a resolution based on similar conditions; if it has worked before in those similar situations, it will likely work in this situation. However, the previous situations may not account for the current circumstances and conditions. Depending on the present condition(s), the resolution may be different.

Rational problem solving allows for a professional to utilize available information to determine an assumption. This may show an optimal solution produced by the specific perception of the professional. However, the rationale would be based on the specific circumstances which may not be relevant to the situation in question; it may provide a minimal view of the situation.

Knowing When to Get Help

It may be unsettling for a safety professional to realize that they may not have sufficient knowledge or skill to handle a particular matter. However, it is imperative that a safety professional recognizes when getting help is necessary. The professional should not feel discouraged by needing help and realize that it is required in certain situations. Sometimes, knowing one's professional limitations shows competency. Additionally, it reduces the likelihood of liability.

First, determine whether help is needed. Sometimes the matter may be resolved by engaging in some research on the matter. This can be done by using professional reports and materials or by conducting a simple internet search. Research will show whether the safety professional may reasonably acquire the necessary knowledge or skill needed to resolve the immediate issue. If this is possible, then the professional should attempt to resolve the matter to the best of their abilities.

If research shows that the safety professional is unable to reasonably acquire the necessary knowledge or skill, then the professional must locate a qualified person. Depending on the issue, a qualified person may be someone that works within the same company or occupational industry. It is best to begin *in house* before attempting to contact external support. Locating qualified support should be based on reasonable beliefs and practices. Before contacting the support, the professional should come up with a potential resolution to the matter and then present their findings. Doing this additional step will increase competency and comprehension. Whether the potential resolution is correct or incorrect, this practice enhances the safety professional's skill. Even if the potential resolution is incorrect, it is best to attempt to understand why it would not work – it increases knowledge for future use.

Practice Test

1. Which process adjustment does NOT align with the philosophy of addressing ergonomic hazards?
   a. Using an adjustable scaffold to change work and/or material height as needed
   b. Selecting the right worker with the skills, strengths, and experience for a particular job
   c. Equipping a delivery truck with a hydraulic lift gate
   d. Fitting job elements to meet the needs of the worker

2. Which of the following provides safety-related input and feedback from diverse key personnel, broadens the reach of safety, and increases safety preplanning effectiveness for better safety and health outcomes?
   a. Management commitment
   b. A good workplace safety culture
   c. Communication policies
   d. Multidisciplinary teams

3. Which of the following can result from safety program incentives in the form of cash awards?

   I. Lower incident rates
   II. Increased awareness
   III. Nonreporting of incidents

   a. I
   b. II
   c. I and II
   d. I, II, and III

4. In heating, ventilation, and air conditioning (HVAC) systems, what do enthalpy wheels do?
   a. Enthalpy wheels primarily function to improve indoor air quality (IAQ).
   b. Enthalpy wheels provide natural ventilation to increase circulation.
   c. Enthalpy wheels transfer heat to incoming air as part of an energy recovery system.
   d. Enthalpy wheels reduce the movement of air particles to reduce the temperature.

5. Which of the following is NOT a benefit of sustainable administrative practices?
   a. Energy conservation
   b. Infrastructure rigidity
   c. Regulatory compliance
   d. Workplace safety

6. Which of the following is a correct statement?
   a. OSHA's safety requirements for aerial lifts can be found in 1926, subpart M, Fall Protection.
   b. Fall protection is required on scaffolds that are six feet or more above the next lower working surface.
   c. Fall protection on aerial lifts is always required, even when less than six feet above the ground.
   d. Stairways and ladders do not require fall protection because OSHA's general duty clause does not apply.

7. How does phytoremediation differ from mycoremediation in terms of treating land pollution?
   a. Mycoremediation can be used for either air pollution or water pollution.
   b. Mycoremediation is a type of ex-situ remediation.
   c. Phytoremediation is used to treat land pollution.
   d. Phytoremediation uses plants to remove contaminants from soil.

8. Which section of the Clean Air Act provides guidance for obtaining permits?
   a. Title II
   b. Title III
   c. Title IV
   d. Title V

9. What is the objective of an incident investigation?
   a. Arrive safely and care for the wounded.
   b. Reduce the contractor's liability obligations.
   c. Identify regulatory and legal violations.
   d. Prevent a similar incident from occurring.

10. What is the most important component for an effective safety program?
   a. Continuous improvement efforts
   b. Responsibility, accountability, and consistent enforcement
   c. Employee involvement
   d. Management commitment

11. Zach is creating a health and safety program for his workplace and wants it to meet the needs of the workforce. Which health and medical conditions need NOT be considered?
   a. Providing health and fitness education
   b. Maintaining minimal communication with an employee's health care provider
   c. Ensuring physical and mental competency of employees
   d. Maintaining healthy work conditions

12. What is the term for the temperature at which a chemical becomes at risk of fire or explosion?
   a. Flammable
   b. Positive pressure
   c. Flash point
   d. Combustible

13. What is the current when a 3.0 V battery is wired across a lightbulb that has a resistance of 6.0 ohms?
   a. 0.5 A
   b. 18.0 A
   c. 0.5 J
   d. 18.0 J

14. Under the Clean Air Act, which pollutant is NOT strictly scrutinized and regulated by the EPA?
   a. Carbon monoxide
   b. Hydrocarbons
   c. Nitrogen dioxide
   d. Particulates

15. Calculate the total load capacity for a deer stand if the allowable compressive stress, F_a, for four steel columns is 4.0 ksi. The cross-sectional area of each column, *A*, is 4.0 in^2.

a. 155 kip
b. 128 kip
c. 64 kip
d. 50 kip

16. What is the greatest barrier to interpersonal communication?

a. Engagement
b. Feedback
c. Noise
d. Tone

17. What is the primary reason green chemistry is considered a sustainable practice?

a. Green chemistry involves cost-effective practices; therefore, production decreases.
b. Green chemistry incorporates recyclable materials in numerous processes.
c. Green chemistry requires less energy than other forms of chemistry.
d. Green chemistry reduces the amount of hazardous waste that's used and produced.

18. A trench is designed with a slope in excess of 45 degrees. The soil in this trench was tested and has a compressive strength greater than 1.5 tsf. Which statement best describes this situation?

a. The trench is in stable rock.
b. The soil in this trench is most likely Type B and dry.
c. The trench cannot be deeper than 10 feet without being designed by a registered engineer.
d. The soil in this trench could be either Type A or stable rock.

19. Which of the following equations is NOT an accurate representation of Ohm's law?

a. $V = I \times R$
b. $R = \frac{V}{I}$
c. $I = \frac{V}{R}$
d. $V = \frac{R}{I}$

20. Which of the following is the study of human physical measurements, physical attributes, and movements within a work environment and provides information that is critical to designing useful ergonomic practices?

a. Anthropology
b. Anthropometry
c. Anthology
d. Health promotion

21. Which of the following is NOT a benefit of feedback?

a. Clear communication
b. Varied communication methods
c. Effective communication
d. Increased performance

22. With regard to confined space activities, which statement about the attendant is NOT true?
   a. The attendant can enter the space only when relieved by another qualified attendant.
   b. The attendant must review the feasibility of rescue equipment.
   c. The attendant must never perform any other activity that interferes with those relevant to confined space duties.
   d. The supervisor can assume the role of attendant.

23. Follow-up procedures for corrective actions include which of the following?
   a. Root cause analysis
   b. "What-if" analysis
   c. "Why?" analysis
   d. Tracking

24. In a workspace atmosphere, a monitoring device's alarm is set to trigger if oxygen levels fall outside what range of percentages?
   a. 21-78%
   b. 12-18%
   c. 10.5-21.5%
   d. 19.5-23.5%

25. Which of the following best identifies the benefits of daily safety walks performed by the safety professional?
   a. Networking with contractors in view of future work opportunities, accomplished by seeking them out during safety walks
   b. Engaging in a meaningful dialogue for safety professionals and workers to find common ground on safety issues
   c. Choosing the proper side in case of an incident so as to favor the side aligned with management
   d. Patrolling the worksite and documenting worker mistakes in the presence of others to set an example

26. A ladder is extended a total of 50 feet in length and positioned to an upper landing. Which of the following are the requirements for this extension ladder?
   a. The ladder must extend three feet above the upper landing with a minimum overlap of three feet.
   b. The ladder must extend four feet above the upper landing with a minimum overlap of three feet.
   c. The ladder must extend three feet above the upper landing with a minimum overlap of four feet.
   d. None of the above

27. Which of the following is most critical for maintaining effectiveness and employee respect for safety programs and policies?
   a. A safety committee comprising key personnel
   b. Management's knowledge of people and processes
   c. A good workplace safety culture
   d. Enforcement

28. What tool is effective for eliminating or reducing communication barriers of a message?
    a. Speaking louder
    b. Feedback
    c. Complex language
    d. One-sided statements

29. Which does NOT apply to lifting?
    a. Keep the load within 10 inches of your body.
    b. Maintain control of the load with your dominant hand.
    c. Keep your back straight, and lift with your legs.
    d. Use a solid, two-handed grip.

30. Who provides important verbal insight about specific tasks during the conduction of a Rapid Upper Limb Assessment?
    a. Equipment manufacturers
    b. Human resources
    c. Supervisors in the facility
    d. The worker being assessed

31. Inspections of machines, materials, and equipment should be led and monitored by which of the following?
    a. The safety professional
    b. Upper management
    c. Qualified engineers
    d. The maintenance department

32. Which of the following does NOT apply to the physical and chemical properties of chemicals?
    a. Flammability
    b. Autoignition temperature
    c. Coagulation
    d. Toxicity

33. Which of the following is an example of nonpoint source water pollution?
    a. Groundwater pollution
    b. Hazardous waste spills
    c. Industrial runoff
    d. Wastewater discharge

34. A rectangular duct has a length of 2 feet and a width of 1.5 feet. If the exhaust rate is measured at 25 feet per minute, what is the ventilation rate?
    a. 12.5 CFM
    b. 25 CFM
    c. 50 CFM
    d. 75 CFM

35. A safety practitioner is assisting in developing controls to address hazards of falls. Several employees are simultaneously at risk. This is a new job, so any system can be used, and elements of feasibility are not an issue. Which system would be most effective for addressing the hazards?

a. Multiple retractable systems consisting of state-of-the-art technology
b. Horizontal lifelines with safety nets as a redundancy
c. Restraints and warning lines in conjunction with a safety monitor
d. Compliant guardrails

36. Which of the following is the most hazardous form of external radiation to a worker?

a. Beta particles
b. Alpha particles
c. Gamma rays
d. Microwave ovens

37. What is the primary way in which work hardening is different from a physical fitness regimen?

a. Work hardening highly emphasizes building mental resilience to work stress.
b. Work hardening includes a nutrition component.
c. Work hardening usually takes place outdoors and is for manual laborers.
d. Work hardening is tailored for a specific person and a specific job task.

38. Who holds the immediate duty to ensure whether a company's employees understand how to reduce workplace risk?

a. Management
b. The associate safety professional
c. The employee
d. The Board of Certified Safety Professionals

39. Which hazard mitigation method is the most proactive?

a. Failure Mode and Effects Analysis
b. Heinrich's domino theory
c. Chain of events theory
d. Fault tree analysis

40. Which measure for the center of a small sample set would be most affected by outliers?

a. Mean
b. Median
c. Mode
d. None of the above

41. Which of the following is NOT a common route of entry for biological and chemical hazards into the body?

a. Absorption
b. Injection
c. Inhalation
d. Respirators

42. Which procedure is an example of in-situ remediation?
    a. Dredging
    b. Extraction
    c. Nanoremediation
    d. Pump-and-treat techniques

43. Workers' compensation is a type of which of the following?
    a. Corporate law
    b. Criminal law
    c. Contract law
    d. Tort law

44. Steve was injured in the workplace and, as a result, he has a broken limb that will have to be amputated. Additionally, he is unable to complete his required duties. What is the classification(s) of Steve's injury?
    a. Partial and temporary
    b. Total and permanent
    c. Total and temporary
    d. Permanent

45. Which of the following is a true statement regarding fall protection on scaffolds in construction?
    a. Fall protection on scaffolds must be provided when they are six or more feet above the ground or the next lower working surface.
    b. Scaffold cross bracing alone may be used in place of a complete scaffold guardrail system in construction.
    c. Employees working below a scaffold do NOT need to wear hard hats when scaffold toe boards are used.
    d. The top rail in a scaffold guardrail system must be installed between 38 and 45 inches from the deck.

46. The energy from electricity results from which of the following?
    a. The atomic structure of matter
    b. The ability to do work
    c. The neutrons in an atom
    d. Conductive materials like metals

47. Regarding emergency planning, which type of event should be given priority for an emergency plan?
    a. Spontaneous events
    b. Regular events
    c. Events that have nominal effect
    d. Casual events

48. Alex is at his workplace and receives a bomb threat over the phone. He is currently on the call. What should Alex do on the call?
    a. Keep the caller on the phone.
    b. Concurrently try to locate the bomb in person.
    c. Make threatening statements.
    d. Tell the caller to quit pranking the company and hang up.

49. When faced with a verbal threat in the workplace, which of the following is a proper response?
   a. Threaten to report the incident to proper workplace personnel.
   b. Make a retaliatory statement.
   c. Report the incident to proper workplace personnel.
   d. Further incite the agitated person.

50. What is the main consideration when preparing for natural disasters?
   a. Timing
   b. Human actions
   c. Location
   d. Types of materials present

51. Mandy is an ASP that is establishing assembly points for persons at a site that handles poisonous gas. What additional factor should she consider in determining an assembly point?
   a. Mobile equipment
   b. Approximate distances between points
   c. Wind conditions
   d. Terrain features

52. Which demonstrates the most effective concept for applying a hierarchy of controls?
   a. Consider personal protective equipment (PPE) requirements for hazards in a new process, and then develop affordable methods of substitution, engineering, and administrative controls.
   b. Assess and eliminate all conceivable hazards in the design stage. Then, continue down the hierarchy, ending with PPE to eliminate all remaining hazards and residual risk.
   c. Assess and eliminate all conceivable hazards in the design stage. Then, assess and minimize the remaining risk to acceptable levels, applying the hierarchy while considering prioritization and feasibility.
   d. Apply the combination of substitution and engineering controls. Then, apply administrative controls and regular employee training as fail-safe redundancies.

53. According to Newton's three laws of motion, which of the following is true?
   a. Two objects cannot exert force on each other without touching.
   b. An object at rest has no inertia.
   c. The weight of an object is the same as the mass of the object.
   d. The weight of an object is equal to the mass of an object multiplied by the acceleration of gravity.

54. Jason wants to establish a disaster recovery plan (DRP). Besides reducing destructive effects, what is another purpose of a DRP?
   a. Cyber security
   b. Recovering essential data
   c. Alternative work locations
   d. Business continuity plan

55. According to the OSHA requirements for stair construction, the stair slope or angle must fall between which of the following?
   a. 20 and 40 degrees
   b. 30 and 40 degrees
   c. 30 and 50 degrees
   d. 40 and 50 degrees

56. Which of the following is performed to determine the root cause of an incident, prevent recurrence, and fulfill legal requirements?
   a. Job safety analysis
   b. Safety audit
   c. Cost-benefit analysis
   d. Incident investigation

57. For a confined space to require a permit, what conditions must be present?
   a. The space must be large enough that a person can fully enter to perform work.
   b. The space must not be intended for permanent or continuous occupancy.
   c. Means of entry and exit must not be limited.
   d. None of the above

58. Which most closely represents the correct sequential flow for a general lockout/tagout (LOTO) procedure?
   a. Notify and shutdown → isolate energy → verify
   b. Identify energy sources → isolate energy → shutdown → verify
   c. Apply LOTO devices → shutdown → isolate residual energy
   d. Relieve stored energy → verify effectiveness → apply LOTO devices

59. Which of the following is NOT a form of asbestos?
   a. Crocidolite
   b. Chrysotile
   c. Amosite
   d. Vermiculite

60. What does PEL mean?
   a. Potential electrical limit
   b. Permissible exposure limit
   c. Protected electrical line
   d. Platform elevation level

61. What are universal precautions?
   a. Sending all personnel on the construction site to receive basic first-aid training, including CPR and AED
   b. Requiring a hepatitis B vaccination for all employees who will be designated first-aid providers
   c. Developing an exposure control plan for all first-aid and medical personnel at a construction site
   d. Treating all bodily fluids as containing blood-borne pathogens or other potentially infectious materials

62. Fiona is conducting a safety and risk assessment of a body shop. She tours the service area and reviews all of the worker service stations. She notices that each station has clearly marked tools and a designated space for each tool. There are one-way pathways for the workers to travel through, and the pathways are free of tools. A storage shed approximately 25 feet from the main service area contains extra gasoline, oil, brake fluid, and other items that may be needed over the course of the workday. In a building adjacent to the service station, there is a customer waiting area and restroom. Fiona finds a door in the customer service area leads to a staff break room. There are lunch items laid out on the counter as well as a toaster oven, blender, and coffeemaker plugged in for staff use. There is also a small kitchen with a microwave, gas stove top, and cooking oils. Next to the stove are several paper towel rolls, sanitizing solution, and bleach. Fiona marks her notes and provides a safety report to the manager. What is one significant issue that Fiona likely assessed?

a. One-way pathways are highly inefficient and would not allow for quick exits during an emergency.
b. The staff break room contains several fire hazards.
c. A single storage shed should only contain one type of mechanical fluid.
d. It is not safe for the customer service area to be connected to the service station.

63. Of the following items, which is NOT used to quantify risk-assessed hazards?

a. Severity
b. Consequence
c. Impact on key performance indicators
d. Likelihood of occurrence

64. What must be established to provide protection to employees, the public, and the environment for potential emergency situations and disasters?

a. Written emergency plan
b. Disaster recovery plan
c. Health and safety programs
d. Hierarchy of needs

65. Barney is a college student who mows lawns for families in his neighborhood. He saves up the money he earns to buy a riding lawnmower to replace his push mower, with the hopes that he will be able to serve more customers in a shorter period of time. He stores his equipment in his parents' garage, which is attached to their home. What is one safety precaution Barney should absolutely take to protect his garage and home?

a. Ensure that his parents do not park their electric cars in the garage.
b. Ensure that all lawnmower gasoline is stored in a traditional red gas can that is tightly sealed.
c. Ensure that all flame sources are 10 feet away from his gasoline containers.
d. Ask one of his customers if he can leave his lawnmower and other equipment on their property.

66. If the half-life of bismuth 212 is 60.5 seconds, with an atomic weight of 211.991 amu, what is the source strength for the decay of one radioisotope of bismuth-212?

a. 3.2111 Bq
b. 0.0115 Bq
c. 0.2021 Bq
d. 0.1461 Bq

67. Which reactive process requires the correct utilization of a series of "Why?" questions?
   a. "What-if" analysis
   b. Root cause analysis
   c. Failure Mode and Effects Analysis
   d. Safety culture analysis

68. Which class of environmental hazard is the Resource Conservation and Recovery Act (RCRA) most closely related to?
   a. Biological
   b. Chemical
   c. Waste
   d. Vermin

69. A sound level meter detects 87 dBA within several construction site areas. These levels exceed any previous readings by 3 dBA. Decibel levels are consistent throughout the course of the normal workday, and earplugs are available. In this situation, the employer will reach compliance through which of the following actions?

> I. Applying effective controls or developing a hearing conservation program (HCP) due to exceeding the action level
> II. Achieving noise attenuation efforts of 3 NRR, to reduce ALL exposures to 84 dBA (below action level), thereby eliminating HCP requirements
> III. Providing hearing protection due to overexposure

   a. I
   b. II
   c. III
   d. II and III

70. Which of the following is a highly effective strategy to reduce musculoskeletal strain?
   a. Clearly marking each person's workspace and ensuring they remain there
   b. Cross-training staff members and regularly rotating job duties and workstations
   c. Offering a weekly employee yoga class
   d. Offering a weekly employee weight training class

71. A facility located near an active tectonic plate boundary develops disaster preparedness plans. Which kind of natural disaster would most likely be included?
   a. Avalanches
   b. Earthquakes
   c. Landslides
   d. Sinkholes

72. Belle works for an arborist company in which she helps with identifying vegetation, providing landscaping suggestions, and removing dead or hazardous trees. After a particularly bad storm in her area, Belle spent approximately five full days of work shifts removing downed or waterlogged trees. She also had to help clear areas of leaves and fallen branches. Using a chainsaw, she had to cut trees into smaller pieces that could be removed; then she used a special tool that lifted any larger pieces into a truck. Finally, she prepared the organic material for recycling and reuse, such as making woodchips. What ergonomic risk was Belle most likely to be affected by during this time?

a. Vibration
b. Posture
c. Repetition
d. Force

73. Kyle is conducting a gap analysis on his company's product production. What is the first step in conducting a gap analysis?

a. Determining gaps within performance
b. Determining policies and procedures to minimize or eliminate gaps
c. Determining current performance
d. Determining desired performance

74. An ASP sees the letter B on a fire extinguisher. Which class of fire does this letter represent?

a. Flammable liquids and gases
b. Electrical equipment
c. Ordinary combustibles
d. Combustible metals

75. Brent is attempting to establish lighting policies using ergonomic principles. If workers in the office with large windows have computer screens that have different lighting levels than the natural light provided in the room, what action should Brent likely take?

a. Require adjustable screen brightness.
b. Purchase blinds to cover the windows.
c. Require protective eyewear.
d. Nothing

76. Todd has entered his workplace to discover that it has been subject to vandalism. What should his top priority be in response?

a. Call government forces.
b. Inform management.
c. Clean up to avoid potential hazards.
d. Touch nothing.

77. Which of the following is NOT a consequence of putting others at risk in the workplace?

a. Risk avoidance
b. Additional training
c. An incident report submission
d. Temporary suspension

78. Michael is a human resources manager at a mid-size hardware company. He needs to fill a position for someone who can help with ordering, organizing, and moving inventory as needed. From an organizational perspective, how can he support workplace ergonomics during the hiring process?
   a. Provide a footrest for the new hire to keep at their desk.
   b. Place the new hire's desk in the inventory room rather than with other workers.
   c. Make sure workers' compensation policies are covered extensively during onboarding.
   d. List necessary physical abilities in the job posting.

79. Which characteristic are adult learners likely to have more of than children?
   a. Creativity
   b. High energy
   c. Internal motivation
   d. Intrinsic intelligence

80. Which safety communication strategy is exclusive to verbal exchanges among all parties involved?
   a. Hand and arm signals
   b. Pictograms
   c. Three-way communication
   d. Presentations

81. A car is traveling at a constant velocity of $25\ m/s$. How long does it take the car to travel 45 kilometers in a straight line?
   a. 1 hour
   b. 3,600 seconds
   c. 1,800 seconds
   d. 900 seconds

82. What is the most eco-friendly way to deal with solid waste?
   a. Composting
   b. Dumping
   c. Recycling
   d. Reusing

83. Which substance or object is an adsorbent solid?
   a. Activated carbon
   b. Cyclone scrubber
   c. Sticky mat
   d. Volatile organic compound

84. Which is NOT a variable that is considered in the NIOSH Lifting Equation?
   a. Location of the item being lifted in relation to the lifter
   b. Level of grip required on the item being lifted
   c. Texture of the item being lifted
   d. Twisting motions that must be made in order to lift the item

85. Which scenario describes a man-made disaster?
   a. A wildfire causes workers to evacuate a worksite.
   b. An engineering issue triggers the collapse of infrastructure.
   c. A hurricane causes a facility to lose electrical power.
   d. An earthquake destroys a laboratory.

86. Which statement regarding incident review processes in construction is true?
   a. When beginning an incident investigation, the first priority is to arrive safely and size up the situation.
   b. A safety administrator is the lead person in most routine incident investigations in construction.
   c. There is typically one causal factor in an incident; the occurrence of multiple causal factors is rare.
   d. Meetings with union representatives must take place before an incident investigation begins.

87. Ergonomics primarily aims to reduce the prevalence and incidence of which health condition?
   a. Obesity
   b. Requiring glasses for nearsightedness
   c. Musculoskeletal disorders
   d. Mood disorders

88. Which of the following does builder's risk cover?
   a. Water damage due to flood
   b. Accidents on the job
   c. Structural damage due to high winds
   d. Claims against the contractor for faulty workmanship

89. What type of test is used to determine whether a chemical is a mutagen?
   a. Defective gene test
   b. Environmental exposure assessment
   c. Ames test
   d. A biopsy

90. Which form of written communication should contain a clearly defined purpose, provide background and facts for the reader, state intentions, finish with a desired outcome, and not exceed two pages in length?
   a. A memo
   b. An email
   c. An executive summary
   d. A lab/field report

91. Which of the following could serve as a feasible work practice control for preschool teachers?
   a. Closing schools each Wednesday and Friday morning
   b. Providing a stand-up desk
   c. Providing an assistant teacher
   d. Providing a 10-minute snack break

92. Stormwater would likely travel efficiently and accumulate the most contaminants by moving across which of the following?
   a. Pavement
   b. Gravel
   c. Undergrowth
   d. Wetlands

93. Which of the following is true regarding silica-associated illnesses and hearing damage?
   a. It needs to be treated within 24 hours of contraction.
   b. It is associated with ingestion-based illness.
   c. It can be prevented but not repaired.
   d. It can be delayed with proper medical therapy.

94. Which of the following is the maximum PEL for worker lead exposure?
   a. 50 micrograms of lead per cubic meter of air during an eight-hour period
   b. 500 micrograms of lead per cubic meter of air during an eight-hour period
   c. 25 micrograms of lead per cubic meter of air during an eight-hour period
   d. 10 micrograms of lead per cubic meter of air during an eight-hour period

95. What is true regarding a personal fall arrest system (PFAS)?
   a. Body belts require that their anchorage supports at least 5,000 pounds per attached worker.
   b. The maximum total fall distance allowed in a personal fall arrest system is six feet.
   c. A PFAS is preferred to fall restraint because the components undergo more reliability testing.
   d. Only components approved by OSHA's product safety division are permitted in fall arrest systems.

96. In an effort to reduce work-related musculoskeletal disorders at the worksite, which of the following ergonomic mitigation factors can be utilized?

I. Redesign tools
II. Modify equipment
III. Use alternate materials
IV. Update work processes

   a. I, IV
   b. II
   c. III
   d. I, II, III, IV

97. To which limits must an organization achieve legal compliance with regard to hazardous substance and chemical exposures?
   a. Threshold limit values (TLVs)
   b. Permissible exposure limits (PELs)
   c. Recommended exposure limits (RELs)
   d. Only PELs and the TLVs that are more stringent than PELs

98. What is the force that opposes motion?
   a. Reactive force
   b. Responsive force
   c. Friction
   d. Momentum

99. Which statement regarding ground-fault current interruption (GFCI) devices is true?
   a. A difference in current between the hot and neutral return wires will cause the GFCI to trip.
   b. The interrupt device will actuate when 70 mA of current or more are detected going to ground.
   c. GFCI is a component of an assured equipment grounding conductor program in construction.
   d. A power cord must have a ground prong (third wire) for GFCI to function as intended.

100. A professional registered engineer must design protection systems if an excavation is greater than how many feet in depth?

a. 10 feet
b. 15 feet
c. 20 feet
d. 25 feet

101. When seeking external help regarding matters beyond a person's current knowledge and skill, what is the first step to solve the immediate issue?

a. Determine the issue.
b. Determine costs of external help.
c. Determine potential external helpers.
d. Determine whether help is needed.

102. Chip handles chemicals at work and is not feeling well. He is experiencing the following symptoms: dizziness, diarrhea, vomiting, and muscle contractions. Of the body portions listed below, which of Chip's body parts or systems have been affected by a symptom?

a. Eyes
b. Chest
c. Nervous system
d. Reproductive system

103. Which of the following assessments is used to identify areas of strength and weakness in the performance of the workforce?

a. Training needs analysis
b. Workforce assessment
c. Organizational assessment
d. Individual assessment

104. Regarding chemical hazard symptoms and factors of risk, which risk factor refers to the amount of chemical in an exposure?

a. Toxicity
b. Dose
c. Personal susceptibility
d. Duration

105. What is an event or series of events with the potential to wreak havoc and/or seriously harm a person or location?

a. Emergency
b. Crisis
c. Disaster
d. Safety program

106. It is a crisp fall day, and Simone is playing on a playground near her home with a friend. She and her friend are running all over the play structure. Each time they slide down the plastic slide, numerous strands of their hair stand straight up. The two friends find this hilarious and continue to keep going down the slide in order to see this phenomenon. What are Simone and her friend experiencing?
a. Static electricity
b. Electrical surges
c. Electrical grounding
d. Electrical impedance

107. In the event of an emergency, what is the priority of things that should be safeguarded?
a. Personnel, processes, and equipment
b. Equipment, processes, and personnel
c. Equipment, personnel, and processes
d. Personnel, equipment, and processes

108. As part of egress from trench excavations deeper than four feet, ladders must be located so that employees have no more than how many feet of lateral travel?
a. 20 feet
b. 25 feet
c. 35 feet
d. 50 feet

109. Which kind of material is considered a biological hazard?
a. Carcinogens
b. Hydrochloric acids
c. Mycotoxins
d. Peroxides

110. Which chemical generally poses the gravest threat to indoor air quality?
a. Argon
b. Carbon dioxide
c. Helium
d. Radon

111. What is the purpose of ethics regarding unethical behavior?
a. Punishment
b. Deterrence
c. Certification revocation
d. Conflicts of interest

112. Which problem solving technique collects information and organizes it into sequential order to determine a resolution?
a. Analytical problem solving
b. Logical problem solving
c. Rational problem solving
d. Creative problem solving

113. As part of fall prevention systems, guardrails must be able to withstand how many pounds of force?
   a. 100 lbs
   b. 200 lbs
   c. 500 lbs
   d. 5,000 lbs

114. Which of the following is NOT a symptom associated with mold exposure?
   a. Respiratory issues
   b. Damaged tissue
   c. Dehydration
   d. Irritated eyes

115. Which of the following is an example of experiential learning?
   a. Independent research assignment
   b. Interactive simulation of tasks
   c. Verbal competency assessment
   d. Written account of relevant experiences

116. Lisa has been working on a new special project at a new location. She has been exposed to a hazardous chemical at the new location. What type of exposure is causing Lisa harm?
   a. Chronic
   b. Acute
   c. Perpetual
   d. Minimal

117. Which of the following is LEAST likely to cause an electrical surge?
   a. Lightning strikes
   b. Plugging numerous appliances in simultaneously
   c. Tripped circuit breakers
   d. Solar flares

118. What is the last course of action a person should take when dealing with a shooting situation?
   a. Confront the shooter
   b. Hide
   c. Run
   d. Contact government forces

119. For hazards of confined spaces, which of the following presents the greatest concern?
   a. Hazardous atmospheres
   b. Lockout/tagout and electrical hazards
   c. Entrapment and entanglement
   d. Blunt force trauma

120. What type of liability imposes an obligation to pay monetary compensation resulting from a lawsuit initiated by a private person or entity?
   a. Civil liability
   b. Tort liability
   c. Criminal liability
   d. Contract liability

121. Control measures used to lower exposure to silica in the workplace may include which of the following?

I. Ventilation
II. Restrictions for time on-site
III. Facial masks with OSHA-approved fine mesh screens
IV. Approved respiration gear

a. II, III, IV
b. I, II, IV
c. II, III
d. I, IV

122. Cathy is experiencing an active shooting in the workplace and has relocated to a hiding location. What should NOT be done in this situation?

a. Lock the doors of the location.
b. Remain silent in the location.
c. Contact government forces.
d. Contact friends and family to inform them of her safety.

123. A virus that continues to replicate within a host cell until the host cell bursts is spreading through which mechanism?

a. Fomite
b. Life cycle
c. Exponential growth
d. Lytic cycle

124. Linda works for a company and has encountered a legal issue at the workplace. She has attempted to understand the issue and solve it, but she and other members of her team are not able to produce a solution. Linda determines that she needs help. Who should help her?

a. Linda's manager
b. A lawyer that is sometimes retained for legal issues
c. The company's general counsel attorney
d. Linda's friend

125. According to the framework established by the National Pollutant Discharge System, which kind of organization is most likely to face additional permit requirements and regulations?

a. Big-box retailers
b. Electronic manufacturers
c. Mining companies
d. Scientific laboratories

126. Under Maslow's hierarchy of needs, what is the foundational need of people?

a. Self-actualization
b. Esteem
c. Love/belonging
d. Physiological

127. Calculate the life cycle cost, using the initial and operational costs, for a $100 25-watt fluorescent bulb that has a life expectancy of 10,000 hours.

a. $25
b. $225
c. $250
d. $55

128. What is the most common form of contact of a bomb threat?

a. Delivered package
b. Phone call
c. Email
d. Statement made in person

129. For a 250 CFM ventilation system, how long will it take to remove air from a room measuring 30 feet wide and 40 feet long with a ceiling height of 10 feet?

a. 0.8 hours
b. 22 minutes
c. 3 hours
d. 80 minutes

130. Which of the following is classified as a trade secret?

a. Suppliers lists
b. Employee addresses
c. Medical records
d. Procurement documents

131. David is concerned with hazards in his workplace and wants to conduct an audit. Which of the following would NOT be used?

a. Site walk around
b. Subjective evaluations
c. Incident reports
d. Employee consultations

132. What would be the most effective engineering control for preventing industrial runoff?

a. Aeration system
b. Bioretention system
c. Hazardous waste manifest system
d. Water spray system

133. What is the sling tension when using a bridle hitch for a steel column that has a total load capacity of 64 kip? The angle between the sling and hitch is 45°, and there are two slings/legs. If the WLL is 75 kip, will the sling break?

a. The sling tension is 85 kip. Yes, the sling will break because 85 kip is greater than 75 kip.
b. The sling tension is 45 kip. No. The sling will not break; 45 kip is less than 75 kip.
c. The sling tension is 75 kip. Yes, the sling will break because 75 kip is equal to 75 kip.
d. The sling tension is 55 kip. No. The sling will not break; 55 kip is less than 75 kip.

134. What is a needs analysis?

a. A needs analysis is a data-gathering activity where all members of an organization are surveyed about specific needs and requests. This information is assessed and then employed in ways to improve the functionality of the organization.
b. A needs analysis is a data-gathering activity where only senior management is surveyed about their specific needs and requests. This information is assessed and then employed in ways to improve the functionality of the organization.
c. A needs analysis is a data-gathering activity where all members of an organization are surveyed about specific needs and requests. However, the surveys must be completed in secret because employees can be fired for expressing critical views of management.
d. A needs analysis is a data-gathering activity only performed at a company after a major safety or security incident.

135. Which of the following is a result of chronic exposure to a mutagenic chemical?

a. A sore throat
b. A spontaneous miscarriage in the second trimester
c. Skin cancer
d. Watery eyes

136. Which of the following is a fall restraint system?

a. Standard guardrail
b. Safety net
c. Shock-absorbing lanyard
d. Full-body harness

137. Following a near miss at the worksite, what should a safety professional do?

a. Admonish the manager of the worksite for failing to supervise and intervene.
b. Deliver a company-wide lecture on the importance of safety.
c. Provide instructional materials to all workers involved in the near miss.
d. Schedule competency assessments if human error was a contributing factor.

138. Which phrase most accurately describes stormwater?

a. Contaminated rainwater
b. Ephemeral surface water
c. Toxic groundwater
d. Untreated wastewater

139. When conducting a competency assessment, what type of assessment is generally preferred because it is more directly related to performance within the industry?

a. Written test
b. Verbal test
c. Practical demonstration
d. Nuanced demonstration

140. What was developed by the BCSP as professional standards for safety professionals to conduct themselves with honesty and integrity?

a. Code of ethics
b. Legal liability
c. Regulatory compliance manual
d. Safety assessments

141. Based on the table of suggested actions and prevention methods for working in the sun, if a worker is stressed at 20 minutes of working in direct sunlight, what is the average WGBT?

WGBT (°F)	Effects when working/exercising in direct sunlight: The body will stress after a specified amount of time (minutes).	Preventive measures: When working/exercising in sunlight, the worker will take a break (in minutes) per hour.
<80	No precautions needed	No precautions needed
80-85	Stressed at 45	15-minute break
85-88	30	30
88-90	20	40
>90	15	45

a. 99 °F
b. 89 °F
c. 85 °F
d. 80 °F

142. Which information-gathering process is used to determine the alignment of goals and attitudes and to assess the overall safety culture of a workforce from top to bottom?

a. Job safety analysis
b. Meeting minutes
c. Survey/questionnaire
d. Data and trend analysis

143. In a chemical equation, the reactants are on which side of the arrow?

a. Right
b. Left
c. Neither right nor left
d. Both right and left

144. Which topics does OSHA require to be addressed in blood-borne pathogens training?

a. Basic first aid, cardiopulmonary resuscitation (CPR), and automated external defibrillator (AED)
b. All relevant provisions of 1926, subpart D, Occupational Health and Environmental Controls
c. Housekeeping, personal protective equipment (PPE), control methods, and hepatitis B vaccine
d. All topics approved by the American Red Cross, Bureau of Mines, or an equivalent curriculum

145. Drivers are involved in proactive defensive training programs and are trained to identify potential hazards. Which of the following should a driver spend the most time reviewing?

a. Defensive driving for vehicles
b. Safe distance procedures
c. Accidents related to mechanical defects and mental health
d. Hazard materials training

146. Which of the following is considered the most effective of controls in ergonomics?

a. Personal protective or support equipment
b. Engineering controls
c. Administrative controls
d. Work practice controls

147. Regarding ISO standards, which standard establishes a framework that an organization may follow to manage environmental effects?

a. 14001
b. OHSAS 18001
c. 19011
d. 45001

148. Which of the following is NOT included in the hierarchy of controls?

a. Administrative
b. Operative
c. Engineering
d. Substitution

149. Which of the following factors is NOT associated with a successful behavior-based safety program?

a. Improved worker communications
b. Improved participation in safety culture
c. Direct contact between workers and OSHA
d. Increased worker information feedback

150. A six-sided die is rolled. What is the probability that the roll is 1 or 2?

a. $\frac{1}{6}$
b. $\frac{1}{4}$
c. $\frac{1}{3}$
d. $\frac{1}{2}$

151. Why are combustible dust explosions especially dangerous?

a. They can set off natural disasters, such as dust storms.
b. They cannot be prevented.
c. They often occur without warning.
d. They can pollute large freshwater resources.

152. Which of the following scenarios is most likely to result in user error?

a. The back-up alert button for a forklift is the same color as the emergency stop.
b. A vehicle does not power on unless the key fob is inside of the car.
c. The night shift in a manufacturing facility plays music while working.
d. A chemical refrigerator is the same model as the lunchroom refrigerator.

153. Ben recently went backpacking through a remote, tropical jungle in a developing country. On the last day of the trip, which was a rather hot day, one of the guides offered Ben a traditional hydrating drink that was native to the culture and the region. Ben enjoyed the drink and participating in the ritual. On his return flight home, he began feeling nauseous, tired, and clammy. By the time his flight landed, he had vomited in the lavatory twice, experienced diarrhea, and had started to feel severe abdominal pain. The airline called for medical help upon landing. What is most likely causing Ben's illness?

a. Jet lag
b. A parasite
c. Heat exhaustion
d. A fungus

154. How do competency assessments relate to training programs?

a. Training programs depend on competency assessments to teach essential skills.
b. Training programs feature competency assessments to provide hands-on experience.
c. Competency assessments provide quantitative data to evaluate training programs.
d. Competency assessments establish an initial baseline for workers' skills.

155. People who are exposed to a chemical fire are also at a high risk of experiencing which of the following scenarios?

a. Electrical fires
b. Downed trees
c. Trauma from shockwaves
d. Tinnitus

156. Which of the following pieces of equipment would help the ergonomics of a desk worker?

a. A rolling chair
b. A monitor with adjustable height
c. A corded desktop telephone
d. A laptop

157. A confined space has the potential to entrap or entangle a worker. Because of this condition, what must the supervisor do in the confined space before work activities begin?

a. Acquire a permit.
b. Ensure that the entrant can't be trapped or entangled.
c. Ensure that the tripod/harness rescue equipment is in place.
d. Practice the work to be performed in an unconfined space.

158. Which of the following is an example of an organizational policy that helps reduce fire hazards in the workplace?

a. Break rooms may only have electrical appliances.
b. Workers must wear safety goggles at all times.
c. Workers must purchase their own safety gear before beginning work.
d. Staff smoking areas are outdoors.

159. Why does on-the-job training have high efficacy?
  a. Workers practice on highly accurate simulators.
  b. Workers independently engage in the learning process.
  c. Workers receive a significant number of written resources.
  d. Workers train with the same equipment they will use every day.

160. What is the main purpose of a gap analysis for a company?
  a. To obtain desired performance
  b. To develop sufficient safety standards
  c. To facilitate crisis management
  d. To ensure sufficient health conditions

161. What is the primary mechanism by which fire blankets eliminate fire hazards?
  a. They remove all oxygen from the fire.
  b. They create a fireproof barrier between the fire and a person.
  c. They protect infants that are unable to leave a building until emergency services arrive.
  d. They drain any fuel sources that are present and stoking the fire.

162. Which is the safest guidance a safety professional can recommend for work involving aerial lifts?
  a. Ensure that guardrails used for fall protection are compliant with 1926 construction safety standards.
  b. Require that all employees in aerial lifts wear a personal fall restraint system when working over water.
  c. Even when body belts are permissible for fall restraint, use a more expensive fall arrest system instead.
  d. Lanyards should be six feet in length since OSHA allows a maximum fall arrest distance of six feet.

163. Which safety process/document has both proactive uses and reactive uses, can be used for training purposes, and shows the breakdown of a larger task into smaller elements with associated hazards and corrective actions?
  a. Incident investigation
  b. Field report
  c. Job safety analysis
  d. Data and trend analysis

164. What is one example of a hazard control strategy that protects the public at large?
  a. Entry and exit personnel documentation
  b. Face shields in a lab
  c. Worker consent forms
  d. Eye wash stations

165. Which of the following is NOT associated with fall protection equipment?

I. Harnesses and lanyards
II. Restraints and guardrails
III. D-Rings and anchor points

a. I
b. II
c. III
d. I and III

166. Which of the following is a potential infection source through the fomite route?

a. Getting coughed on by a person who has the flu
b. Eating a sandwich that was prepared by someone who did not wash their hands
c. Touching a doorknob that someone sneezed on
d. Watching a movie with someone who has active herpes simplex type 2

167. Which of the following settings is most at risk of experiencing a chemical fire?

a. A residential home with an electric cooktop
b. A manufacturing facility that makes lawn fertilizers
c. A manufacturing facility that makes car body parts
d. A studio that practices heavy sound engineering

168. Fires that result from the presence of metals such as magnesium or potassium require a different approach by firefighters. Suppose that a Class D fire occurs in an outside environment. Which of the following substances or strategies would be useful to extinguish these types of fires?

a. Sand/dirt
b. Water
c. Reducing the oxygen concentration
d. Smothering the fire with a cotton blanket

169. Nathan is creating a list of crisis response items to ensure that he has the necessary items to respond to an emergency. Which of the following would likely NOT be considered proper response equipment?

a. An eye wash station
b. Candles
c. Batteries
d. A crisis response plan

170. Which of the following effects of extreme heat causes permanent tissue damage?

a. Heat rash
b. Heat stroke
c. Both of these
d. Neither of these

171. A compounding factor of a trench collapse can include which of the following?

a. The use of timber instead of aluminum in shoring methods
b. Dense soils at greater depths
c. Not using a qualified engineer with excavations deeper than 10 feet
d. Increased depth and the presence of water

172. What is the main purpose of a cost-benefit analysis?
   a. When evaluating a policy or program, a cost-benefit analysis empirically tests its efficacy to ensure that resources aren't squandered.
   b. Cost-benefit analyses are rarely conducted because they are expensive and unreliable.
   c. When evaluating a policy or program, a cost-benefit analysis is conducted that rationally tests its efficacy to ensure that resources aren't squandered. However, because cost-benefit analyses are antiquated, management typically decides on the policy or program based on its organizational popularity.
   d. A cost-benefit analysis is the empirical testing of a policy or program. However, management is typically disdainful of them because of a belief that the testers are inherently biased.

173. What is the difference between lagging and leading indicators?
   a. Lagging indicators show areas where an organization is falling behind performance goals, while leading indicators show areas where an organization is pulling ahead of performance goals.
   b. Lagging indicators highlight an organization's weaknesses compared to others in the industry, while leading indicators indicate its relative strengths.
   c. Lagging indicators reflect performances that happened in the past, while leading indicators reflect activity that can change future performance or success.
   d. Lagging indicators show where an organization's processes are outdated or obsolete, while leading indicators show where an organization has effectively implemented cutting-edge processes and technology.

174. Rachel was injured in the workplace and, as a result, she is only able to conduct a portion of the duties required by her position. What is the classification of her injury?
   a. Total
   b. Permanent
   c. Partial
   d. Temporary

175. A dosimeter has a readout of 20%. Calculate the estimated exposure if a worker is wearing earmuffs with an NRR of 25 dBA and round to the nearest integer.
   a. 106 dBA
   b. 78 dBA
   c. 60 dBA
   d. 18 dBA

176. A chain of command is selected for reacting to emergency situations. These individuals should be selected based upon which of the following?
   a. Years of experience with the organization
   b. Level of title, i.e., beginning with upper management
   c. An individual's capacity to perform under stress
   d. Job responsibilities

177. How does the BCSP Code of Ethics require professionals to avoid conflicts of interest?
   a. Avoid practices that could deceive workers or the public.
   b. Balance the interests of the client, property owner, employees, and public.
   c. Conduct relationships with professional integrity.
   d. Present only truthful information about their qualifications.

178. Which of the following statements is correct?
   a. Only a competent person may approve the removal of a ground prong from an electrical cord.
   b. A continuous path to ground is NOT necessary for powered tools that are GFCI-protected.
   c. Equipment manufactured prior to 1971 is exempt from OSHA's electrical safety standards.
   d. The use of tools that are NOT grounded is permissible so long as they are double-insulated.

179. Elizabeth is acquiring samples for a consolidated report. Regarding ethical principles, what matters should be given consideration in the report?
   a. Preferential sampling
   b. Minimal sampling
   c. Objective sampling
   d. Exaggerated sampling

180. What is one informal way to conduct an exposure assessment in the workplace?
   a. Schedule a quarterly appointment with a third-party auditor.
   b. Ask an administrator if any exposures were reported recently.
   c. Create a process workflow.
   d. Review a process checklist during the morning huddle.

181. What is most commonly the biggest barrier to extinguishing a fire?
   a. Multi-head sprinkler systems, which are now obsolete
   b. Reaching the fire in time
   c. Removing the oxygen source
   d. Removing the fuel source

182. Of the following, which is NOT a site inspection employed by OSHA?
   a. Imminent danger
   b. Fatality/catastrophe
   c. Supplemental inspection
   d. Follow-up

183. Mina is conducting a professional development satisfaction survey. She needs to survey 75% of her organization, which has over 3,000 employees. What is the best way for her to collect detailed data from this many people?
   a. Online survey
   b. Paper survey
   c. A series of focus groups
   d. Face-to-face interviews

184. Before submitting to an audiometric test, how long must an employee avoid exposure to excessive noise?
   a. 8 hours
   b. 12 hours
   c. 14 hours
   d. 24 hours

185. Which type of control includes worker training or the selection of specific personnel for a task?
   a. Substitution
   b. Engineering
   c. Administrative
   d. Personal protective equipment

186. Which skill is most beneficial for interpersonal communication?
   a. Active listening
   b. Commercial expertise
   c. Professional experience
   d. Technological literacy

187. Hard hats are vital for protecting against electrical shocks, bumps, and impacts. OSHA standards have adopted the hard hat testing and rating criteria of which entity?
   a. American Society of Testing and Materials (ASTM)
   b. National Institute of Occupational Safety and Health (NIOSH)
   c. American National Standards Institute (ANSI)
   d. Underwriters Laboratories (UL)

188. Regarding hazardous chemicals, what is the name for the exposure limit that must not be exceeded at any time?
   a. Top limit
   b. Tier limit
   c. Base limit
   d. Ceiling limit

189. Which of the following is one of the most common sources for infections?
   a. Soil
   b. Food
   c. Cleanrooms
   d. Anterooms

190. Which form of written communication should omit fine details, provide data and results, and not exceed two pages in length?
   a. A memo
   b. An email
   c. An executive summary
   d. A lab/field report

191. What is true of emergency planning?
   a. Notifying emergency management personnel is always the first action taken in an emergency.
   b. Employer emergency action plans (EAPs) are addressed by OSHA's HAZWOPER standard.
   c. The first priority in an emergency is to safeguard personnel, equipment, and processes.
   d. Emergency alarms in construction must have a sustained sound level of at least 105 A-weighted decibels.

192. Janet is developing a presentation for her workplace regarding newly implemented procedures at the workplace. She wants to effectively communicate to approximately 10 recipients. What is the best form of communication for Janet to use for the presentation?

a. Speech
b. Newsletter
c. Meeting
d. Phone conference

193. With regard to workforce diversity and a multicultural public, which form of communication would be most effective in communicating a jobsite hazard?

a. Written
b. Oral
c. Signage
d. Email

194. Which is true of fire extinguishers in an outdoor construction environment?

a. OSHA requires fire extinguishers on construction sites to be inspected at least once per week.
b. *Dry chemical* and *multipurpose dry chemical* can be said to refer to the same thing.
c. An extinguisher must be located within 10 feet of five or more gallons of flammable materials.
d. There must be one fire extinguisher for every 10 employees assigned to a construction site.

195. What would be an effective method of manipulating powdered materials?

a. Barricades
b. Locked cabinets
c. Glove box system
d. Sealed system

196. In which of the following scenarios would a thermal detector be useful in preventing fire emergencies?

a. Directly near a window in a home
b. In a supermarket
c. Directly near a natural gas pipe in one's home
d. In a large residential home

197. Which of the following hazard assessment methods uses frequency and severity to determine the level of hazard risk? Resultant risk levels can be qualitative or quantitative.

a. Failure Mode and Effects Analysis
b. "Why?" methodology
c. Risk matrices
d. Job safety analysis

198. Janine is a 32-year-old woman who manages a production facility where medical devices are made. One step of the production line involves lifting a part and using a motorized hand tool to tighten a joint before the device can be sent for quality assurance. One day, Janine learns that two of her production technicians are out sick and are expected to miss 10 days of work. One production technician is a woman who is considerably taller and visibly more muscular than Janine. The other technician is an 18-year-old man who is about the same stature as Janine. Their absences will delay production significantly, so Janine and two other colleagues offer to fill in during shifts. What is a work process that may change as Janine fills in on this production line?

a. Janine will likely not be able to fill this role.
b. Janine may need to use the motorized hand tool for shorter durations of time.
c. Janine will need to hire new staff immediately.
d. Janine will need to ask the colleagues to provide all lifting support.

199. Concerning BCSP ethics and writing, how should a safety professional prepare a written statement or opinion of a work site's condition?

a. Subjective and competent
b. Honest and biased
c. Unambiguous and objective
d. Clear and disproportionate

200. Sparks can fly up to how many feet?

a. 45 feet
b. 50 feet
c. 35 feet
d. 40 feet

Answer Explanations

1. B: Outdated work methods to maximize production were aligned with fitting the worker to the job. This philosophy maximized repetition and increased the risk to workers. The modern philosophy of ergonomics is to fit the job to the needs of the worker, as evidenced in Choices *A*, *C*, and *D*.

2. D: Multidisciplinary teams focus on how safety relates to multiple departments and their respective processes. Management commitment is critical for leadership and supporting a multitude of program elements, processes, and policies. A good safety culture is helpful for the respect of safety and the performance of safe practices. Communication is vital for relaying information. Although Choices *A*, *B*, and *C* are all important, they are not specific to safety across multiple departments.

3. D: While increased awareness of potential hazards is a benefit of safety programs with cash incentives, over time, employees can begin to interpret cash bonuses as part of their regular income. In the event of an incident, employees will often conceal the event, if possible, to avoid what they perceive as a cut in pay for themselves as well as others. This results in ostensible lower incident rates, but the true result is loss of control. Alternative elements, such as recognition awards (e.g., coats, hats) or gift cards, can serve to curb nonreporting of incidents.

4. C: HVAC systems use enthalpy wheels to recover energy by transferring heat to the incoming air. As a result, the HVAC system doesn't require as much energy to provide cool air. Thus, Choice *C* is the correct answer. HVAC systems are commonly used to improve indoor air quality (IAQ), but enthalpy wheels aren't directly related to this function. Therefore, Choice *A* is incorrect. Unlike HVAC systems, natural ventilation leverages natural phenomenon to improve airflow, like through a window opening. Therefore, Choice *B* is incorrect. Enthalpy wheels play a role in heating, not cooling, so Choice *D* is incorrect.

5. B: The question is asking for something that is NOT a benefit of sustainable administrative practices. Energy conservation means reducing the consumption of energy, like using natural ventilation or turning off machines when they aren't being used, which is a sustainable practice. Therefore, Choice *A* is incorrect. The government sometimes requires industries to implement sustainable policies and practices, so regulatory compliance can be considered a benefit of sustainability. Therefore, Choice *C* is incorrect. Choice *D* is incorrect because some sustainable polices, like green chemistry, involve a reduction in hazardous waste, which improves workplace safety. Infrastructure rigidity indicates an unwillingness to adapt existing infrastructure or depart from prior conventions, which is the opposite of transitioning to green energy or other sustainable practices. Thus, Choice *B* is the correct answer.

6. C: Fall protection, typically a personal fall arrest system (PFAS), is always required when working on aerial lifts; the lift platform does not have to be six or more feet from the ground. Although not treated as scaffolds, OSHA's aerial lift requirements can also be found in 1926, subpart L, Scaffolds. OSHA does not require that fall protection be provided on scaffolds until they are 10 feet above the ground or the next lower working surface, though it should be provided whenever feasible to do so. OSHA's general duty clause has nothing to do with specific stairway and ladder safety requirements; the general duty clause is a stipulation that an employer can be cited for an employee hazard exposure even when there is no specific standard that addresses the particular hazard in question.

7. D: Phytoremediation and mycoremediation are both in-situ methods of remediation used to mitigate land pollution, especially in terms of removing heavy metals from soil. Phytoremediation involves using plants to remove contaminants, while mycoremediation extracts contaminants with fungi. Thus, Choice

D is the correct answer. Mycoremediation isn't typically used to treat or mitigate air pollution, so Choice *A* is incorrect. Choice *B* is incorrect because mycoremediation is a type of in-situ remediation. Phytoremediation and mycoremediation are both used to treat land pollution, so Choice *C* doesn't describe a difference and must therefore be incorrect.

8. D: The Clean Air Act provides guidance for obtaining permits under Title V. The permitting system is national and overseen by the EPA; state and local agencies are heavily involved in the issuance and enforcement of permits. Thus, Choice *D* is the correct answer. Title II establishes emission standards for vehicles and aircrafts, so Choice *A* is incorrect. Title III features a comprehensive list of hazardous air pollutants along with emission standards; therefore, Choice *B* is incorrect. Choice *C* is incorrect because Title IV is focused on the regulation of noise pollution.

9. D: The objective of an incident investigation is to prevent a similar incident from occurring. Arriving at an incident site in a safe manner and caring for the wounded must take place before the fact-finding begins, but these actions are not objectives of the investigation itself. Liability, regulatory, and legal matters will always be topics of concern during an incident investigation process, but they are of secondary importance to the primary goal of ensuring the incident does not repeat itself.

10. D: Management commitment must be in place. Management develops and maintains programs, grants approvals for corrective actions and associated costs, and has the most control over elements contained within Choices *A, B,* and *C,* among many others.

11. B: Choice *B* is correct because the program should support open communication with an employee's health care provider. Choice *A* is incorrect because education offers a source of information to an employee regarding their health and fitness. Choice *C* is incorrect because ensuring competency shows that an employee is fit to perform their required duties. Choice *D* is incorrect because maintaining healthy work conditions decreases an employee's potential harm in the workplace.

12. C: A substance's flash point is the temperature at which it releases vapors that are flammable or combustible. Highly volatile chemicals have low flash points, while more stable chemicals have higher flash points. The other terms listed do not describe this specific term, so Choices *A, B,* and *D* are incorrect.

13. A: According to Ohm's law: $V = IR$, so using the given variables:

$$3.0\ V = I \times 6.0\ \Omega$$

Solving for I:

$$I = \frac{3.0\ V}{6.0\ \Omega}$$

$$I = 0.5\ A$$

Choice *B* shows a miscalculation in the equation by multiplying, rather than dividing, 3.0 V by 6.0 Ω. Choices *C* and *D* are labeled with the wrong units; joules measure energy, not current.

14. B: The Clean Air Act is an incredibly comprehensive regulatory statute, and it places especially strict criteria on the emission of carbon monoxide, Choice *A;* nitrogen dioxide, Choice *C;* and particulate matter, Choice *D*. The other primary targets of the Clean Air Act are lead, ozone, fine particulate matter, and sulfur dioxide. Hydrocarbons are commonly used to produce fuel and chemical products. Methane is a hydrocarbon and greenhouse gas; however, it isn't regulated to the same extent as the pollutants discussed above.

15. C: First, calculate the load capacity for one steel column.

$$P_{load} = F_a A = 4.0 \frac{klb}{in^2} \times 4.0\ in^2 = 16\ klb$$

Now calculate the total load capacity for all four steel columns.

$$Total\ load\ capacity\ XE\ " = \sum_{n}^{\#\ columns} P_{load};\ n = \sum_{n}^{4} P_{load};$$

$$n = 16\ klb + 16\ klb + 16\ klb + 16\ klb$$

$$= (4)16\ klb = 64\ klb\ or\ kip$$

16. C: Interpersonal communication involves multiple parties delivering messages and feedback about those messages. Noise can be defined as anything that distracts or disrupts a message, such as flowery language or broken syntax. Therefore, noise undermines interpersonal communication because it functions as a barrier to communication. Thus, Choice *C* is the correct answer. Engagement refers to the amount of interaction or level of response a message is likely to receive, and engagement is almost always positive. Therefore, Choice *A* is incorrect. Choice *B* is incorrect because feedback is a component of effective interpersonal communication. Tone is related to the emotional connotation and denotation of a message, and it can either be positive or negative. So, Choice *D* is incorrect.

17. D: Green chemistry is a sustainable practice because it involves reducing the use and production of hazardous materials, usually through substitution or the selection of an alternative design. Consequently, green chemistry can improve workplace safety and diminish the risk of pollution. Thus, Choice *D* is the correct answer. Green chemistry isn't always cost-effective, and Choice *A* doesn't specify which type of production would decrease. As a result, Choice *A* is incorrect. Green chemistry can involve recyclable materials, but this doesn't describe its primary feature or purpose. So, Choice *B* is incorrect. Likewise, green chemistry could reduce energy consumption in some scenarios, but this isn't necessarily true or the reason it is considered sustainable. Therefore, Choice *C* is incorrect.

18. D: Type A and stable rock have a higher compressive strength (> 1.5 tsf), which gives these soil types better cohesive properties. Higher cohesion results in a soil's ability to support itself under greater pressure. Type B soil's compressive strength does not exceed 1.5 tsf and is deemed unstable due to fissuring or having been previously disturbed. Less stable soils are less self-supportive and more likely to collapse with greater slope, lateral pressure, and trench depth. Type A soils can be sloped up to 53 degrees, and stable rock can be vertical.

19. D: Ohm's law describes the relation between voltage and amperage, where voltage is a measure of electromotive potential much like potential energy of motion in kinetics. Amperage measures the electric current or the flow rate of 1 coulomb of electrons in a second. The basic law is $V = I \times R$, but this can be manipulated in the following ways:

$$I = \frac{V}{R} \quad or \quad V = I \times R \quad or \quad R = \frac{V}{I}$$

20. B: Anthropometry is the scientific study of the human body's measurements and proportions and incorporates how these affect interactions with work environments and tools. Anthropology is the study of human societies and cultures, so Choice *A* is incorrect. Anthology refers to a collection of written prose or music, so Choice *C* is incorrect. Health promotion is a field that focuses on individual or community health. While aspects of this may be used in the workplace, it focuses more on social and behavioral determinants rather than workplace influences, so Choice *D* can be eliminated.

21. B: Choice *B* is correct because although feedback can be used with various communication methods, variety is not a benefit of feedback. Choice *A* is incorrect because feedback allows for clear communication between the speaker and listener. Choice *C* is incorrect because feedback attempts to provide effective communication for all parties involved. Choice *D* is incorrect because feedback may allow for increased performance in a workplace.

22. B: The attendant is NOT responsible for retrieval systems and feasibility. This responsibility belongs to the supervisor. With training, the supervisor can also act as the attendant or entrant. The attendant has many responsibilities that are vital to safety for confined space activities. The attendant must be fully dedicated to these duties and cannot perform any other duty unless relieved by another qualified attendant.

23. D: Corrective actions must be tracked and monitored in order to determine the measure of effectiveness and to determine if additional or alternative actions are required. Analysis techniques, such as root cause, "Why?," and "what-if," are used to determine what corrective actions may be appropriate.

24. D: Oxygen levels are approximately 21% in normal air/atmosphere. Lower oxygen levels result in negative effects on human health. Oxygen levels outside the normal range also affect LEL and UEL with respect to combustion and explosion.

25. B: Choice *B* is the sole answer associated with the safety and impartiality an ASP must exercise on the worksite.

26. D: Ladders must extend no less than three feet above an upper landing. In addition, extension ladders up to 36 feet require a minimum overlap of three feet, while those between 36 and 48 feet require at least four feet of overlap. Extension ladders between 48 and 60 feet (such as this ladder that extends a total of 50 feet) require a minimum overlap of five feet. Therefore, none of the answers are correct.

27. D: Employees will not take safety seriously without equal and consistent enforcement of safety and policy violations. Although effective safety committees, management knowledge, and positive safety cultures play critical roles in safety, without equal and consistent enforcement practices, safety programs, policies, and procedures are rendered meaningless.

28. B: Choice *B* is correct because feedback reduces communication barriers and allows for better understanding. Choice *A* is incorrect because speaking louder generally does not reduce confusion with a message's content. Choice *C* is incorrect because complex and technical language may lead to confusion based on their potentially limited knowledge. Choice *D* is incorrect because effective communication requires both the speaker and the listener to engage in discussion to provide a clear message understood by all parties.

29. B: Keeping the load close to the body, maintaining a straight back, and using the two-handed grip are all part of sound lifting technique. The dominant hand is immaterial when lifting loads. The back is at its strongest when straight and all things—including bending the legs, maneuvering for proximity to the load, and body positioning—are done to ensure the back remains straight regardless of the body position.

30. D: A Rapid Upper Limb Assessment accounts for subjective feedback provided by the worker who is being assessed. The worker provides information about what tasks they believe may be higher risk than others and works with an external evaluator to ultimately assess these risks. Equipment manufacturers, Choice *A*, may provide information about how to use equipment safely, such as through manuals or initial set-up, but they are not consistently in a facility or working with different staff members during use. Human resources, Choice *B*, typically does not play a role in on-the-floor assessments. Supervisors, Choice *C*, may provide some input or change work processes based on the outcomes of an assessment, but they are not key stakeholders during the actual assessment process.

31. A: A key element of inspections and preventive maintenance efforts is safety. Documentation of inspections, relative scheduling, safety controls applied, and tracking and monitoring for verification of performance level are reasons inspections should be led by the safety professional, although both upper management and qualified engineers might be involved in the inspections. The maintenance department is often unqualified to conduct safety inspections.

32. C: Flammability and toxicity are associative chemical hazards. Autoignition temperature is a physical property. Flammability is a substance's likeliness to ignite or burn and cause combustion or fire. Toxicity is the term associated with human exposure to a toxin or poison (acute toxicity means a significant exposure over a short time period). Autoignition temperature is the minimum temperature at which a chemical or other substance can ignite without an external ignition. Coagulation is merely a result of blood desiccation.

33. C: Nonpoint source water pollution is the accumulation of pollutants from a variety of sources. Industrial runoff is considered nonpoint source water pollution because the water isn't being contaminated directly from a distinct source. Instead, industrial runoff involves rainwater accumulating pollutants and transferring them to a body of water. Thus, Choice *C* is the correct answer. Groundwater pollution frequently occurs as a result of point source pollution, such as an accident involving an underground storage tank. Therefore, Choice *A* is incorrect. Hazardous waste spills are an example of point source water pollution because the spill is the direct source of pollution. Similarly, wastewater discharge is more strongly associated with point source pollution since the discharge is a direct source.

34. D: The formula for the ventilation rate must account for a rectangular duct.

$$Q = VA = V \times (length \times width)$$

$$= \frac{25\,ft}{min} \times 2.0\,ft \times 1.5\,ft = 75\,\frac{ft^3}{min}\ or\ 75\ CFM$$

35. D: Among the choices, only guardrails can prevent the fall from occurring. All of the other options address the hazard after the fall or have elements that can be defeated. Fall arrest systems are designed to prevent the worker from hitting the ground but can still result in injury.

36. C: Microwave ovens, alpha particles, and beta particles denote non-ionized radioactivity, whereas gamma rays are ionized radiation, which is potentially very harmful to humans.

37. D: Work hardening is a type of program that takes place after a musculoskeletal injury has occurred, with the goal of strengthening a worker to perform their specific job responsibilities. Each work hardening regimen will vary based on the worker's role. It is not often focused on mental resiliency or nutrition; instead, it focuses more on strength and biomechanics. Therefore, Choices *A* and *B* can be eliminated. Work hardening can take place for all jobs, not just outdoor or physically laborious ones, so Choice *C* can be eliminated.

38. A: Choice *A* is correct because management has the initial duty to ensure whether an employee understands how to reduce risk in the workplace. Choice *B* is incorrect because it is not the duty of the associate safety professional to ensure whether a company's employee understands how to reduce risk. Choice *C* is incorrect because management is obligated to ensure its employees can safely conduct their jobs. Choice *D* is incorrect because the Board of Certified Safety Professionals certifies that licensed safety professionals are qualified and develops relevant rules and regulations for safety professionals.

39. A: FMEA involves prevention, pre-planning analysis, and other efforts relative to safety through design. The domino and chain of events theories are "after the fact" measures that apply corrective actions to mitigate risk.

40. A: The mean would be most affected by outliers. An outlier is a data value that's either far above or below the majority of values in a sample set. The mean is the average of all values in the set. In a small sample, a very high or low number could greatly change the average. The median is the middle value when arranged from lowest to highest. Outliers would have no more of an effect on the median than any other value. Mode is the value that repeats most often in a set. Assuming that the same outlier doesn't repeat, outliers would have no effect on the mode of a sample set.

41. D: Respirators are actually a form of protective equipment that help in purifying air that is inhaled, so they work to prevent entry of biohazards. Making contact with hazards through the eyes and skin (absorption), puncturing the skin to allow entry of hazards (injection), breathing in hazards (inhalation), and consuming hazards (ingestion) are the most common ways that biological and chemical hazards get into the body, so Choices *A, B,* and *C* are incorrect.

42. C: In-situ remediation refers to the on-site treatment of contaminants, meaning the soil and water isn't removed. Nanoremediation is a type of in-situ remediation because it involves the introduction of nano-sized agents to absorb contaminants in soil or water. Thus, Choice *C* is the correct answer. Choice *B* is incorrect because extraction is the removal of soil for treatment at an off-site facility. Similarly, dredging involves the extraction of soil in bodies of water, so Choice *A* is incorrect. Pump-and-treat techniques remove contaminated water and send it through a filtration system. In addition, other varieties of pump-and-treat systems include transportation to a treatment facility. Therefore, Choice *D* is incorrect.

43. D: Choice *D* is correct because workers' compensation may be viewed as a type of tort law. Choice *A* is incorrect because corporate law generally deals with matters of corporate procedures and formal compliance. Choice *B* is incorrect because criminal law involves criminal violations and proceedings initiated by a government entity. Choice *C* is incorrect because contract law involves agreements between two or more people.

44. B: Choice *B* is correct because Steve is unable to complete the required duties and will suffer from the injury for the remainder of his life. Choice *A* is incorrect because Steve's injury will not allow the full completion of required duties. Choice *C* is incorrect because it is impossible to fully recover from an amputated limb; it is not a temporary injury. Choice *D* is incorrect because the classification of Steve's injury may also be total.

45. D: OSHA requires that the top rail of a standard scaffold guardrail system be installed between 38 and 45 inches from the deck. For other horizontal working surfaces in construction, such as a rooftop, the top rail must be installed between 39 and 45 inches. Fall protection on scaffolds is not mandatory until they are more than 10 feet above the ground or the next lower working surface. OSHA's general rule to provide fall protection at six feet in construction does not apply to scaffolds. OSHA allows a scaffold's cross brace to take the place of one, but not both, of the guardrails. Employees working below others should not forego the use of hard hats when toe boards are used on scaffolds because toe boards will not prevent all objects from falling; hard hats are mandatory whenever there is a falling-object hazard.

46. A: The physical structure of the atoms that compose matter lends itself to the production of electricity. The arrangement of the subatomic particles and the associated charges—mainly the negatively charged electrons in the cloud—are associated with the ability to create an electric current, which can be harnessed to do work.

47. B: Choice *B* is correct because regular events are events that are most likely to occur and should be given priority in an emergency plan because they are more certain. Choice *A* is incorrect because spontaneous events are not as likely to occur. Choice *C* is incorrect because priority should be given to events that have more than a nominal effect—the bigger the event's impact, the more priority the event is given. Choice *D* is incorrect because casual events are similar to spontaneous events.

48. A: Choice *A* is correct because Alex should keep the caller on the phone and gather as many facts as possible to relay to government authorities. Choice *B* is incorrect because attempting to locate the bomb should be left to government authorities. Choice *C* is incorrect because making threatening statements may escalate an already dangerous situation. Choice *D* is incorrect because bomb threats should be taken seriously and handled in an appropriate manner.

49. C: Choice *C* is correct because verbal threats should be handled by proper workplace personnel. Choice *A* is incorrect because a threat to report the incident may escalate the situation in that moment. Choice *B* is incorrect because a retaliatory statement may incite the agitated person to execute a verbal threat. Choice *D* is incorrect because incitement will likely escalate the situation and lead to harm.

50. C: Choice *C* is correct because the location of a facility is most likely to determine what type of natural disaster can be expected. Choices *A*, *B*, and *D* are incorrect because they are more often associated with man-made disasters.

51. C: Choice *C* is correct because wind conditions may carry poisonous gases to persons depending on their location related to wind conditions; this requires an additional consideration. Choice *A* is incorrect because mobility is relative to how freely groups of people may move. Choice *B* is incorrect because the amount of distance shows how quickly a person may reach an assembly point. Choice *D* is incorrect because terrain features, again, show how freely groups of people may move.

52. C: Choice *C* represents the soundest concept among the choices. All of the options suggest some amount of remaining risk. For Choice *A*, PPE should be viewed as a last resort, and cost alone should never rule the selection of controls. For Choice *B*, inherent risk can be mitigated, but it cannot be eliminated. For Choice *D*, with no controls in design or pre-planning, using a combination of (or multiple) controls suggests a high presence of armed hazards inherent within the process.

53. D: According to Newton's second law of motion, $F = m \times a$. Weight is the force resulting from a given situation, so the mass of the object needs to be multiplied by the acceleration of gravity on Earth: $W = m \times g$. Choice *A* is incorrect because, according to Newton's first law, all objects exert some force on each other, which is based on their distance from each other and their masses. This is seen in planets, which affect each other's paths and those of their moons. Choice *B* is incorrect because an object in motion or at rest can have inertia; inertia is the resistance of a physical object to change its state of motion. Choice *C* is incorrect because the mass of an object is a measurement of how much substance there is to the object, while the weight of an object is gravity's effect on the mass.

54. B: Choice *B* is correct because recovering essential data is another purpose of a DRP. Choice *A* is incorrect because cyber security is a growing concern of how to safeguard a DRP, but it is not the purpose of the DRP. Choice *C* is incorrect because alternative work locations may be part of a DRP, but it is not the DRP's purpose. Choice *D* is incorrect because a business continuity plan is distinct from the DRP, but both may work concurrently.

55. C: OSHA 1926.1052(a)(2) states that stair slope or angle must fall between 30 and 50 degrees from horizontal. Steeper or higher angles result in tread overlap or require a diminished tread depth. Such elements reduce the safety for traversing stairs. Higher and lower angles represent components of ladders and ramps, respectively.

56. D: Other processes are performed for many reasons. The performance of a JSA, safety audit, and CBA are not indicative of a reaction to an incident. Audits test performance and compliance to programs and standards. A CBA is primarily a spending based initiative, and a JSA is performed initially to prescribe procedures and detect associated job hazards. Determination of root cause, prevention of recurrence, and fulfilling legal requirements all point to a process directly reactive to an incident. The results of an incident investigation could, however, spur the actuation of a JSA revision, exterior (OSHA) audit, or CBA as the extension of an improvement/recommendation expenditure.

57. D: These are the characteristics that identify a confined space. A permit-required confined space must contain one of several possible hazards, including hazardous atmospheres, engulfment, entanglement, and electrical hazards. Confined spaces present situations where workers are closely bound to potential hazards. Close proximity results in more easily triggered events that are difficult to isolate and evade. Confined spaces also magnify hazard risk level and duration of exposure.

58. A: Proper notification to affected personnel and machine shutdown are vital to initiating a lockout/tagout (LOTO) procedure. Effectiveness must be verified last via try-out.

59. D: Crocidolite, chrysotile, and amosite are forms of asbestos, and vermiculite is not. Used in coatings, cements, and insulators for high-heat environments such as steam engines, crocidolite is blue and has the highest heat resistance. Chrysotile is found in floors, walls, roofs, and ceilings of residential and commercial edifices and is the most common form of asbestos. It is also found in brake linings, gaskets, seals, and insulation. Amosite is amphibole-type asbestos, as it has needle-like fibers. Known as brown asbestos, it originates mainly in Africa. Used in pipe insulation and cement, it is also found in ceiling tiles and insulation boards with high asbestos content (up to 40%). Vermiculite is a hydrous silicate material that expands when heated, although it is not a form of asbestos.

60. B: The permissible exposure limit (PEL) is a standard and defining term utilized by OSHA to define worksite exposure limits to a substance in the air over a given span of time. Time-weighted average (TWA) is the exposure to any hazardous substance on the worksite over an 8-hour workday/40-hour week, which if exceeded, mandates the worker exit the worksite. Short-term exposure limit (STEL) allows for 15-minute exposure increments (four times daily) if TWA is not exceeded; if exceeded, workers must exit the worksite for one hour. The rest of the answer choices are nonsensical terms.

61. D: *Universal precautions* refers to the practice of treating all bodily fluids as if they contain blood-borne pathogens, which is planning for the worst. This is a highly recommended practice since it is impossible to know whether blood or other potentially infectious materials contain pathogens without access to employee medical data, which a first-aid responder will typically not have, and also because an employee may have a blood-borne disease and be unaware of it.

62. B: The staff room is highly cluttered and full of disorganized items, and there are flammable materials near the gas stove. It also seems like there are too many appliances plugged in near each other. These are all fire hazards. One-way pathways can promote efficiency and streamlined workflows; emergency escape routes do not necessarily need to follow workflow pathways, so Choice *A* can be eliminated. Some mechanical fluids can be stored near each other if they do not have incompatible reactivities and are stored safely, so Choice *C* can be eliminated. There is no reason the customer service waiting area cannot be adjacent to the service area, especially if enough barriers (e.g., tempered glass, additional separation) are present, so Choice *D* can be eliminated.

63. C: Severity, consequence, and likelihood of occurrence are all quantifiable items for assessed hazards. Impact on key performance indicators is more of a corporate concern.

64. A: Choice *A* is correct because the written emergency plan establishes a plan to provide public, environmental, and employee protection for emergencies and disasters. Choice *B* is incorrect because a disaster recovery plan is a set of procedures established to address only a disaster situation. Choice *C* is incorrect because health and safety programs are created to protect the health and safety of employees and make up philosophy of management. Choice *D* is incorrect because the hierarchy of needs shows the basic needs of a person, not how to provide protections in the event of emergency situations or disasters.

65. B: Gasoline for personal or business use, such as for lawn equipment, should be stored in conventional red containers that are tightly sealed. This is to prevent the release of vapors, which are flammable and can spread if they catch on fire. If Barney's parents' cars can fit in their garage, the electric vehicles do not necessarily pose a risk by being parked near Barney's equipment, so Choice *A* can be eliminated. Gasoline containers should be placed a minimum of 50 feet away from any flame sources, not 10, so Choice *C* is incorrect. It is not necessary, or professional, for Barney to ask customers to store his equipment if he has a clean and secure storage area, such as a garage, so Choice *D* is incorrect.

66. B: First, calculate the decay constant for bismuth-212:

$$\lambda = \frac{0.693}{60.5}\ s = 0.0115\ s$$

Next, calculate the source strength. The atomic mass of bismuth-212 is 212 atomic mass units (amu), which includes a specific number of protons and neutrons. Convert atomic mass units to grams using the following conversion unit.

$$M = 212\ amu \times \frac{1.66054 \times 10^{-24}\ g}{1\ amu} = 3.52 \times 10^{-22}\ g$$

The atomic weight of bismuth is $W = 211.991$ amu or 211.991 grams/(gram-atom). The gram-atom is equivalent to a mole. Therefore, the source strength for the decay of bismuth-212 is:

$$Q = \frac{\lambda M N_A}{W} = \frac{(0.0115\ s^{-1})(3.52 \times 10^{-22}\ g)(6.022 \times 10^{23}\ atoms \times (g - atom)^{-1})}{211.991\ g \times (g - atom)^{-1}}$$

$$Q = 0.0115\ Bq$$

67. B: The "Why?" method is used as part of incident investigations to determine the root cause. Incident investigations are reactive processes, meaning an event has already occurred (Why did this happen?). "What-if" analysis is a preplanning method that questions what could go wrong (What could happen?). Failure mode and effect analysis applies mathematical modeling to determine potential rates of failure and reliability for systems such as materials and equipment. A safety culture analysis can be conducted via surveys, interviews, observations, data, etc., for the primary purpose of promoting shared responsibilities and safe work practices.

68. C: The Resource Conservation and Recovery Act (RCRA) is the primary basis for most of the regulations related to waste disposal, including municipal, industrial, and hazardous waste. Thus, Choice *C* is the correct answer. The RCRA addresses biological hazards and chemical hazards, but the focus is on treating, storing, and disposing these when they are categorized as hazardous waste. Therefore, Choices *A* and *B* are both incorrect. Vermin are animals that threaten human health and/or facilities. The RCRA is primarily concerned with waste, not animal control, so Choice *D* is incorrect.

69. A: This situation suggests that an HCP is not in place. These levels may not necessitate anything more than effective controls to reduce levels at the source. The information given does not suggest overexposure in an eight-hour workday (over 90 dBA), and NRR devices should not be applied to environmental levels (outside of the ear).

70. B: Cross-training employees so that they can rotate tasks and workstations reduces physical repetition, both in the short term and the long run. Repetition is a high risk factor for developing musculoskeletal disorders. Additionally, this supports general organizational output by ensuring that workers can fill in for others as needed. Therefore, this is an effective solution on multiple levels. Choice *A* is incorrect because ensuring that people complete the same tasks day after day in the same space places a worker at high risk for injury. Offering yoga and fitness classes is a nice benefit that can help physically and mentally strengthen a worker; however, there is no guarantee that workers will use the classes, and one weekly class is likely not enough frequency to create resilience. Therefore, Choices *C* and *D* can be eliminated.

71. B: The movement of tectonic plates releases energy from the lithosphere, and the resulting seismic waves cause earthquakes. Facilities located near tectonic plate boundaries are at a significantly higher risk of earthquakes than other areas. So, in this scenario, the facility's disaster plans would heavily feature earthquake preparedness. Thus, Choice *B* is the correct answer. Avalanches involve the rapid movement of snow down a hill or mountain. While earthquakes can trigger avalanches, there are numerous other causes, including precipitation, wind, and animals. In addition, the question doesn't indicate whether the facility is located near a snow-capped mountain. As a result, Choice *A* is incorrect. Similarly, earthquakes can cause landslides, but the question doesn't mention a slope at risk of collapsing. So, Choice *C* is incorrect. Sinkholes are formed when chemical processes cause the Earth's surface layer to collapse, which isn't directly related to plate tectonics. Therefore, Choice *D* is incorrect.

72. A: Utilizing chainsaws for extended periods of time causes a vibration risk that can lead to musculoskeletal issues. The case indicates that Belle worked in a number of different environments with different tree issues, so she was likely not repeating the same series of events nor holding the same posture, so Choices *B* and *C* can be eliminated. She also had specialized lifting tools to help her with larger pieces of wood, so force was likely not as large of an issue as vibration was during the storm clean-up. Therefore, Choice *D* is incorrect.

73. C: Choice *C* is correct because determining current performance is the first step in a gap analysis. The steps of a gap analysis are: (1) determining current performance; (2) determining desired performance; (3) determining gaps within performance; and (4) determining policies and procedures to minimize or eliminate gaps. Therefore, Choices *A*, *B*, and *D* are incorrect.

74. A: Class B fires involve flammable liquids and gases. Ordinary combustible fires are Class A, electrical equipment fires are Class C, and Class D fires involve combustible metals.

75. A: Choice *A* is correct because adjusting screen brightness allows for even light levels at nominal cost. Choice *B* is incorrect because purchasing blinds will require more resources to be used and would not be as cost effective as adjusting screen brightness levels. Choice *C* is incorrect because protective eyewear is likely unnecessary and would require more cost than adjusting screen brightness. Choice *D* is incorrect because the extended lighting imbalance could cause harm to eyes and increase the likelihood of fatigue.

76. D: Choice *D* is correct because Todd should refrain from touching anything so that the scene is preserved for law enforcement personnel to evaluate and engage in fact finding. Choice *A* is incorrect because this is not the top priority when confronted with a scene of vandalism. Choice *B* is incorrect because law enforcement personnel should be contacted prior to management. Choice *C* is incorrect because cleanup should not occur until after law enforcement personnel have evaluated the scene.

77. A: Choice *A* is correct because an employee who practices risk avoidance will not knowingly put others at risk. Choice *B* is incorrect because putting others at risk in the workplace may require an employee to engage in additional training regarding risk avoidance and safety practices. Choice *C* is incorrect because putting others at risk in the workplace may lead to an incident report being added to the employee's personnel file. Choice *D* is incorrect because temporary suspension is a possible consequence that could be used to deter employees from putting others at risk.

78. D: By listing physical expectations in the job description (e.g., must be able to lift 50 pounds, must be able to sit for four hours at a computer), Michael is communicating the company's needs and ensuring that potential applicants are informed of the physical requirements of the job. This supports workplace ergonomics by trying to prevent a poor fit between an applicant and a job responsibility that could later lead to injury. Footrests, Choice *A*, can support workers whose feet do not reach the floor when sitting at a desk. However, they may not be necessary for all workers, and this is not a type of HR policy. Isolating a worker to remain alone in a warehouse may not be good for their psychological health at work; they may also be in the way of other workers who are managing products, so Choice *B* can be eliminated. Ergonomics aims to reduce or eliminate workers' compensation claims by promoting workplace health and safety and generally reducing risk to the worker. Emphasizing this aspect of a worker's job benefits is not necessarily a good prevention and risk reduction strategy. Therefore, Choice *C* is incorrect.

79. C: Adult learning theory leverages the skills and preferences of adults, which are very different from children's skills and preferences. Compared to children, adults are much more likely to have internal motivation, which is why adult learning theory emphasizes independence and self-direction. Thus, Choice *C* is the correct answer. While adults can be creative, Choice *A*, and possess intrinsic intelligence, Choice *D*, both characteristics can also be held by children. Therefore, Choices *A* and *D* are both incorrect. Children are more likely to have high energy than adults, so Choice *B* is incorrect.

80. C: This communication method increases process safety through awareness. Three-way communication is exclusive to verbal command and the responses of those involved in or near to the activity. Hand and arm signals are used where interference and deficiencies exist in line-of-sight or audible detection. Pictograms are visual modes to increase comprehension.

81. C: The answer is 1,800 seconds:

$$\frac{(Desired\ Distance\ in\ km\ \times\ conversion\ factor\ (m\ to\ km))}{current\ XE\ "Current"\ \ velocity\ in \frac{m}{s}}$$

$$\frac{\left(45\ km\ \times\ \frac{1{,}000\ m}{km}\right)}{25 \frac{m}{s}} = 1{,}800\ seconds$$

82. D: Alongside reducing consumption, reusing materials is considered to be the most eco-friendly practice in terms of solid waste since reusing doesn't generate significant waste. Composting is a sustainable practice since it effectively recycles organic compounds and improves the fertility of soil; however, non-organic components of the products are generally thrown away, so composting is considered less eco-friendly than reusing materials. Similarly, recycling involves repurposing materials, but it is not as efficient as reusing entire products. Therefore, Choice *C* is incorrect. Choice *B* is incorrect because dumping solid waste into landfills is the least environmentally friendly practice due to the risks it poses to groundwater and land.

83. A: Adsorption is the attraction of atoms and molecules to a surface, and activated carbon is one of the most commonly used solid adsorbents. For example, activated carbon can be used to purify air, clean up spills, and remediate groundwater. Thus, Choice *A* is the correct answer. Cyclone scrubbers are used to remove air contaminants by swirling gas around a chamber to remove particles, so Choice *B* is incorrect. Sticky mats are used to segregate particulates and to prevent the spread of contamination. Therefore, Choice *C* is incorrect. Choice *D* is incorrect because volatile organic compounds are fragrant organic chemicals and air pollutants.

84. C: The texture of the item being lifted is not considered in the NIOSH Lifting Equation, which is a validated measure that is used to assess lifting risks of various items. The item location in relation to the lifter, grip required, and any twisting motions are considered in the Lifting Equation, so Choices *A, B,* and *D* are not the correct answers. Additionally, the Lifting Equation considers the frequency of lifting of the item and the distance that the object is being lifted.

85. B: Man-made disasters differ from natural ones in that man-made disasters are directly caused by human actions. An engineering issue is a human action, and it directly contributed to the collapse of infrastructure, which is a disaster. Thus, Choice *B* is the correct answer. Choice *A* is incorrect because it doesn't mention how the wildfire started, so it can be safely assumed to be a natural disaster. Choices *C* and *D* are both incorrect because hurricanes and earthquakes are classic examples of natural disasters.

86. A: When beginning an incident investigation, the priorities are to arrive safely, size up the situation, care for the injured, and lastly, protect the property. A safety administrator should not be the lead person in routine incident investigations; the most immediate supervisor, such as a supervisor, is the lead person in most routine incident investigations. There are frequently multiple causal factors in an incident. Actions such as corrective measures, disciplinary proceedings, and meetings with union representatives should only take place once useful factual data has been gathered.

87. C: The primary focus of ergonomics in workplace settings is to reduce and eliminate musculoskeletal disorders and injuries by eliminating repetitive motion and improper weight loads. Its primary focus is not to address additional health conditions, such as obesity, eye problems, or mood disorders, although these can arise from poor ergonomics or as a byproduct of musculoskeletal disorders. Choices *A, B,* and *D* can be eliminated.

88. C: Builder's risk covers damage to the property, equipment, or materials due to natural disasters or during normal work, but it does not cover floods, Choice *A*, or faulty workmanship, Choice *D*. General liability covers accidents on the job, Choice *B*, and is a separate policy from builder's risk.

89. C: Choice *C* is correct because an Ames test is used to determine whether a chemical is a mutagen. Choice *A* is incorrect because a defective gene test is not the actual name of a known test. Choice *B* is incorrect because an environmental exposure assessment would relate to environmental factors. Choice *D* is incorrect because a biopsy is not a test to determine whether a chemical is a mutagen.

90. A: A memo is used to introduce or lead actions that require subsequent attention. The content is not complex enough for report form and is too large for email.

91. C: Providing an assistant teacher can inherently create job rotation and reduce some of the physical strain that preschool teachers face (e.g., picking up children, squatting down to help with hands-on learning, cleaning, sanitizing). Although closing schools for a period during the week would provide a nice break, it is not feasible for continuous learning; therefore, Choice *A* can be eliminated. A preschool teacher is not a traditional desk worker and likely would not benefit from having a stand-up desk, so Choice *B* can be eliminated. Finally, one 10-minute snack break does not provide enough break or task rotation for a worker, so Choice *D* can be eliminated.

92. A: Stormwater travels the fastest on smooth, paved surfaces, such as roadways and parking lots. In addition, the lack of soil and shrubbery means there are more contaminants on the pavement for the rainwater to carry. Thus, Choice *A* is the correct answer. Stormwater would flow through gravel and accumulate substantial contaminants; however, stormwater would move more efficiently across pavement because there are no barriers obstructing the stormwater's path. Therefore, Choice *B* is incorrect. Choices *C* and *D* are incorrect because the construction of wetlands and planting of undergrowth are both strategies used to remove contaminants from stormwater.

93. C: Many illnesses and injuries in the workplace/elsewhere can be treated and/or cured. Silica-associated illnesses and hearing damage (if allowed to occur) are simply irreversible once established.

94. A: Fifty micrograms of lead per cubic meter of air during an eight-hour period defines the OSHA-prescribed exposure limit; the remaining answers are wrong.

95. B: The maximum total fall distance allowed in a personal fall arrest system is six feet; this includes all attached components and is not based on the length of the lanyard alone. A PFAS must utilize a full-body harness; body belts are never allowed with fall arrest systems. Fall restraint is always preferred to fall arrest since no fall can occur. However, sometimes restraint systems are not possible, and the contractor has no choice and must employ a PFAS. PFAS components must meet OSHA's strength requirements outlined in 1926, subpart M, but OSHA itself does not perform product safety testing or approve any person, place, or thing.

96. D: All answers entail applicable methods designed to provide safeguards against work-related injuries. The redesign of tools counteracts the forces of torque, weight, and impact on the worker and improves tool usability. Equipment modifications for the worksite may include noise reduction or installment of guards to prevent worker injury. Alternating materials are concurrent with attempts to replace materials containing lead, silica, asbestos, and other harmful chemicals with less harmful ones. The manner in which a job is physically done can invariably be improved. Every task should be reviewed and evolved to better suit the ergonomic and physical environment of a worker.

97. B: Threshold limit values and recommended exposure limits are both recommendations and are NOT enforceable. An employer can, however, choose to protect workers with limits (TLVs and RELs) more stringent than OSHA PELs.

98. C: The force that opposes motion is called friction. It also provides the resistance necessary for walking, running, braking, etc. In order for something to slide down a ramp, it must be acted upon by a force stronger than that of friction. Choices *A* and *B* are not actual terms, and Choice *D* is the measure of mass multiplied by velocity ($p = mv$).

99. A: GFCI will interrupt current when a disparity of 5-7 mA is detected between the hot and neutral return wires and will take place within 0.025 seconds (1/40 of a second) when functioning properly. An assured equipment grounding conductor program is an exhaustive system of documentation and testing that is the only alternative to using GFCI in construction. A power cord does not need a ground prong for GFCI to function correctly, since the ground wire plays no part in the operation of the GFCI, so even non-grounded power tools that are double insulated can be protected by GFCI.

100. C: Excavations greater than 20 feet require a professional registered engineer. OSHA 1926.652, 1926 Subpart P App B, and other parts express the requirements for registered engineers with respect to protective systems. Complexities in required design are proportional to trench depth. This means that as trench depth increases, the compounding dynamics of the soil's weight compounds the lateral pressure forces that test a soil's compressive strength in contrast to trenches of lesser depth. Increased depth also increases the potential for the presence of water, which results in soil fissuring, compromising sidewall integrity. Trench failure at such depths requires ensured effectiveness in design that exceeds the capabilities of a non-engineer.

101. D: Choice *D* is correct because determining whether help is needed is the first step in deciding whether external help should be sought regarding an issue. Choice *A* is incorrect because at this point, the issue has already been determined. Choice *B* is incorrect because determining costs of external help is a subsequent step to determining whether help is needed. Choice *C* is incorrect because determining potential helpers would be completed after determining whether help is needed (and likely concurrent with determining costs of external help).

102. C: Choice *C* is correct because muscle contractions are a symptom related to the nervous system. Chip is also experiencing dizziness (head, not listed) and diarrhea and vomiting (stomach, not listed). Choice *A* is incorrect because Chip is not showing symptoms of irritated or red eyes. Choice *B* is incorrect because Chip is not experiencing any trouble or complications regarding his ability to breathe. Choice *D* is incorrect because Chip is not experiencing any symptoms related to gametes (egg or sperm).

103. D: During an individual assessment, leaders assess groups and individuals to determine which skills they have and which skills they need to perform at the desired level determined by the organizational assessment, Choice *C*. The training needs analysis, Choice *A*, determines how to deliver the training to the workforce based on their limitations. The workforce assessment, Choice *B*, refers to the individual assessment.

104. B: Choice *B* is correct because dose refers to the amount of chemical experienced in an exposure. Choice *A* is incorrect because toxicity refers to a chemical's toxicity level, not the amount of the chemical. Choice *C* is incorrect because personal susceptibility refers to the likelihood of a person experiencing a chemical symptom based on that person's specific characteristics. Choice *D* is incorrect because duration refers to the amount of time that a person experiences a chemical exposure.

105. C: Choice *C* is correct because an event or series of events with the potential to wreak havoc and/or seriously harm a person or location is the definition of a disaster. Choice *A* is incorrect because an emergency is a spontaneous situation that requires an immediate response. Choice *B* is incorrect because a crisis is a spontaneous event that poses a threat to an organization and requires time to act. Choice *D* is incorrect because a safety program is a developed system implemented by an organization of safety policies and procedures to minimize loss from harm.

106. A: Simone and her friend are experiencing static electricity, which occurs when two unbalanced objects experience friction against one another. The loss of electrons from their bodies to the slide is causing their hair to have a slightly positive charge, and each individual hair begins to repel the others. This is not an electric surge, Choice *B*, which is a massive burst of electrons. Electrical grounding, Choice *C*, occurs when electricity is dispersed in a safe manner. This normally occurs by routing electricity back to the earth or a grid designed to disperse excess electrons. This is not what is happening with the two friends. Electrical impedance, Choice *D*, is a measure of electricity within a circuit with resistance and is not relevant here.

107. D: Choice *D* is correct because the order of priority is personnel, equipment, and processes. Choices *A*, *B*, and *C* are incorrect because they do not reflect the correct order of priority.

108. B: Per 1926.651 for safe means of egress (reduced distance of lateral travel in any direction), 25 feet of lateral travel is the maximum distance permissible by ladders. This is true for stairs, ramps, or any other means used in trenches greater than four feet deep.

109. C: Mycotoxins are a toxic metabolite produced by certain fungus, including many species of mold. Since mycotoxins are produced by a living organism, they are classified as a biological hazard. Thus, Choice *C* is the correct answer. A carcinogen is anything that plays a role in causing the formation and spread of cancer. Biological hazards can be carcinogenic, but carcinogens are much broader. For example, many chemical hazards are carcinogens. Therefore, Choice *A* is incorrect. Hydrochloric acid is found in many animals' digestive systems, and it is also used as an industrial-strength chemical in laboratory settings. So, hydrochloric acid is classified as a chemical hazard, and Choice *B* is incorrect. Likewise, Choice *D* is incorrect because peroxides are chemical compounds, with the most commonly used being hydrogen peroxide.

110. D: Along with carbon monoxide and asbestos, radon poses one of the most serious threats to indoor air quality (IAQ), and it is a leading cause of lung cancer in the United States. Thus, Choice *D* is the correct answer. Argon is one of the most abundant atmospheric gases, and it is relatively harmless to humans. Therefore, Choice *A* is incorrect. The human body produces carbon dioxide, and it has a negligible effect on IAQ unless concentration is extremely high. Therefore, Choice *B* is incorrect. Helium is non-toxic and only a threat to human health when directly inhaled from a container. So, Choice *C* is incorrect.

111. B: Choice *B* is correct because the purpose of ethics is to deter unethical behavior. Choice *A* is incorrect because punishment should be reserved for perpetual or extreme offenses/violations. Choice *C* is incorrect because certification revocation would be a punishment for extreme conduct. Choice *D* is incorrect because ethics attempts to eliminate conflicts of interest to preserve integrity.

112. B: Choice *B* is correct because logical problem solving compiles prior information and data and organizes it into sequential order to determine a resolution based on similar circumstances. Choice *A* is incorrect because analytical problem solving evaluates objective conditions and aspects of a situation. Choice *C* is incorrect because this is a problem-solving technique based on an associate's specific perception. Choice *D* is incorrect because this is a problem-solving technique using potential outcomes of a situation to determine a resolution, but it involves a willingness to encounter risk.

113. B: Guardrails must be able to withstand 200 pounds of force. This applies to a side force applied within two inches of the top, and a down force applied to the top of the guardrail, per OSHA 1926.502.

114. C: Choice *C* is correct because mold exposure does not generally cause dehydration. Choice *A* is incorrect because mold exposure causes respiratory issues, such as wheezing and coughing. Choice *B* is incorrect because mold may harm persons with immune system deficiencies and cause damage to tissue. Choice *D* is incorrect because mold exposure may cause irritated eyes.

115. B: Experiential learning involves providing workers with hands-on and/or real-world learning opportunities. The interactive simulation of tasks is an example of experiential learning because it allows workers to actively solve problems in a hands-on context. Thus, Choice *B* is the correct answer. An independent research assignment would be consistent with adult learning theory, but it isn't experiential learning since it doesn't involve learning by doing. Therefore, Choice *A* is incorrect. Verbal competency assessments can evaluate whether a worker has the necessary knowledge or skills, but answering questions isn't comparable to hands-on practice. So, Choice *C* is incorrect. Similarly, Choice *D* is incorrect because writing about relevant experiences isn't interactive.

116. B: Choice *B* is correct because Lisa has been working on a new project at a new location. Choice *A* is incorrect because the fact that these are new circumstances shows that it has not been long enough for the exposure to be classified as chronic. Choice *C* is incorrect because the term is *chronic exposure*, and there has not been sufficient time for this classification to apply. Choice *D* is incorrect because the term is *acute exposure*, not *minimal exposure*.

117. D: Solar flares are geomagnetic storms that erupt and cause bursts of radiation that shoot out from the Sun. Solar flares that are intense enough to affect the Earth's electrical grid are not impossible, but they are extremely rare. Therefore, this is the least likely cause of electrical surges. Lightning strikes, appliance overuse, and tripped circuit breakers are far more common causes that can occur daily. Therefore, Choices *A, B,* and *C* are incorrect.

118. A: Choice *A* is correct because pursuing a shooter should be the last course of action and should only be done when a person's life is in imminent danger. Choice *B* is incorrect because attempted hiding should occur prior to shooter confrontation. Choice *C* is incorrect because this should be the first attempted course of action. Choice *D* is incorrect because contacting government forces should be a priority during a shooting situation.

119. A: Although all choices represent hazards, air quality has been (and continues to be) the greatest hazard for confined space workers.

120. A: Choice *A* is correct because civil liability imposes an obligation to pay monetary compensation resulting from a private lawsuit. Choice *B* is incorrect because tort liability is a type of civil liability, but it is not the only type of liability that requires monetary compensation. Choice *C* is incorrect because criminal liability results from a criminal proceeding initiated by a government entity. Choice *D* is incorrect because contract liability is a type of civil liability, but it is not the only type of liability that requires monetary compensation.

121. B: Ventilation, in concert with testing and monitoring, is an inexpensive and effective way to purge an environment of hazardous airborne particles. Time on-site controls, including PEL, TWA, and STEL, are to be utilized when the work locale cannot be safely purged of most suspended particles. Particulate respirators are the least expensive but do not protect against vapors, gases, or chemicals. Respirators work by filtering, purifying, or utilizing exterior air. Other respirators, such as cartridge and gas mask respirators, are used to filter or purify specific substances.

122. D: Choice *D* is correct because Cathy should remain silent until it is safe for her to reveal her location or until government forces instruct otherwise. Choice *A* is incorrect because Cathy should lock the doors of the location. Choice *B* is incorrect because Cathy should remain silent so that a shooter will not hear her. Choice *C* is incorrect because Cathy should contact law enforcement forces so that they may assist.

123. D: In the lytic cycle, a virus infects a host cell and utilizes the cell's functionality to reproduce itself. It makes copies until the cell cannot fit any more viral particles within its cell wall, consequently causing the cell to burst (or lyse). Copies of the virus leak out of the host cell, infect nearby host cells, and continue the process to spread throughout a region. A fomite is a surface, such as a door or countertop, that may have infectious material on it. It is not a cyclic mechanism through which a virus spreads; therefore, Choice *A* can be eliminated. The life cycle refers to an organism's journey from conception to death. A virus is not a living organism, and it does not undergo this process to spread, so Choice *B* can be eliminated. Exponential growth refers to a doubling of an entity. While highly contagious viruses can spread at this rate, it is not a term that describes the replication mechanism, so Choice *C* can be eliminated.

124. C: Choice *C* is correct because Linda is faced with a legal issue, an attorney is trained on legal matters, and the general counsel is considered *in house*. Choice *A* is incorrect because the facts state that no one on Linda's team may produce a solution, and the manager is likely on Linda's team. Choice *B* is incorrect because the lawyer is considered external because of the sporadic retention, and the lawyer is not the general counsel. There are not enough facts to determine whether the general counsel may not address the matter, making Choice *C* the better choice. Choice *D* is incorrect because there are no facts to show that Linda's friend is competent regarding the subject matter and is likely not a good choice.

125. C: As part of its regulatory authority, the National Pollutant Discharge System provides permits to businesses for discharging pollutants and using water. Some industries face more stringent requirements due to the increased amount of hazards in their wastewater, like mining, large-scale agriculture, and the oil and gas industry. Thus, Choice *C* is the correct answer. Big-box retailers, Choice *A*; electronic manufacturers, Choice *B*; and scientific laboratories, Choice *C*, all generate considerable waste, but mining faces more stringent regulations due to the inherent toxicity and sheer amount of chemicals involved in the mining process.

126. D: Choice *D* is correct because Maslow's hierarchy of needs are the following, in order: (1) physiological; (2) safety; (3) love/belonging; (4) esteem; and (5) self-actualization. Therefore, Choices *A*, *B*, and *C* are incorrect because of the hierarchy outlined above.

127. D: First, calculate the initial costs given.

$$Ci = cost\ of\ asset \times \left(\frac{usage\ hours}{life\ expectancy}\right) = \$25 \times \left(\frac{10{,}000\ hr}{10{,}000\ hr}\right) = \$25$$

Now, calculate the operational costs. A 25-watt light bulb uses 25 watts per hour; it will consume 1 kilowatt per hour (kWh) if operated for 10 hours.

The number of hours to use 1 kWh is equal to:

$$Hours\ to\ use\ 1\ kWh = \frac{1{,}000\ W}{25\frac{W}{hr}} = 40\ hr$$

Now, find the operational costs.

$$Co = cost\ of\ 1\ kWh\ \times \left(\frac{usage\ hours}{hours\ to\ use\ 1\ kWh}\right) = \$0.12 \times \left(\frac{10{,}000\ hr}{40\ hr}\right) = \$30$$

The total life cycle costs for the fluorescent bulb are:

$$LCC_{f\ bulb} = Ci + Co = \$25 + \$30 = \$55$$

The life cycle cost of an incandescent light bulb was found to be:

$$LCC_{bulb} = Ci + Co = \$230 + \$120 = \$350$$

128. B: Choice *B* is correct because bomb threats are usually communicated through a phone call on the company's landline. Choice *A* is incorrect because a delivered package may be a bomb threat, but it is not the primary communication of a bomb threat. Choice *C* is incorrect because an email is not a common form of communication for a bomb threat. Choice *D* is incorrect because it is not common for a bomb threat to be made in person.

129. A: To find the time in hours, use the following equation.

$$Air\ change\ per\ hour\ (ACH) = \frac{60\ min}{1\ hr} \times \frac{Q}{area\ \times ceiling\ height}$$

The ACH should be equal to 1 because one cycle of air will be removed from the room.

$$\frac{1}{t} = \frac{60\ min}{1\ hr} \times \frac{Q}{A = area\ \times ceiling\ height}$$

$$Time = \frac{area \times ceiling\ \times height}{\frac{60\ min}{1\ hr} \times Q} = \frac{30\ ft \times 40\ ft \times 10\ ft}{\frac{60\ min}{1\ hr} \times 250\ \frac{ft^3}{min}} = 0.8\ hr$$

130. A: Suppliers lists, customer lists, blueprints, and product specs are examples of trade secrets. Procurement documents, Choice *D*, are considered only confidential information. Employee addresses, Choice *B*, are personally identifiable information, and medical records, Choice *C*, are confidential data protected by HIPAA.

131. B: Choice *B* is correct because an audit must be conducted on objective evaluations, not subjective evaluations, that may be corroborated. Choice *A* is incorrect because a site walk around allows for an objective perception of a workplace. Choice *C* is incorrect because incident reports may show hazards that have previously occurred and their frequency. Choice *D* is incorrect because employee consultations (interviews) allow for additional insights based on experiences at a workplace.

132. B: Industrial runoff is created when rainwater accumulates pollutants and travels to a body of water. Bioretention systems use organisms to absorb contaminants, which prevents the contaminants from entering runoff. For example, a row of native plants is a bioretention system that can be used to remove contaminants and limit industrial runoff. Thus, Choice *B* is the correct answer. Aeration systems are most commonly used in wastewater and soil to break down contaminants. While aeration systems would limit industrial runoff, bioretention systems are more focused on preventing floods and runoff, so Choice *A* is incorrect. Choice *C* is incorrect because the hazardous waste manifest system is used to regulate the transportation of hazardous waste. Choice *D* is incorrect because a water spray system is used to segregate particulates and prevent contamination.

133. B: Use the sling tension equation to find its value in kip.

$$Sling\ XE\ "\ tension = \frac{load\ (kip)}{\#\ legs \times sin(\theta)};\quad \theta = angle\ between\ sling\ and\ load$$

$$Sling\ XE\ "\ tension = \frac{64\ kip}{2 \times sin(45°)} = 45.25\ kip$$

Because the sling tension is less than the working load limit (WLL), the sling will not break.

134. A: A needs analysis is the process in which an organization gathers information about the principal needs and requests of its members. This analysis studies the expectations and requirements of subjects who are affected by workplace programs or regulations. Surveys and questionnaires are the primary means of procuring information about specific needs and requests. After the needs analysis is conducted, the information is assessed and incorporated into plans to improve the functionality of the organization.

135. C: Mutagenic agents cause genetic changes through errors in DNA replication or other cellular mutations. While most mutagens result in cancer and are considered carcinogenic, some simply cause damage beyond regular cell functioning that cannot be remedied. A sore throat and watery eyes are acute, rather than chronic, responses to irritants, so Choices *A* and *D* can be eliminated. When an exposure causes poor fetal outcomes, it is known as a teratogen rather than a mutagen, so Choice *B* can be eliminated.

136. A: Standard guardrail systems are considered a fall restraint because they prevent a fall from happening. A safety net is an arrest system. A shock-absorbing lanyard and full-body harness are both components of a personal fall arrest system (PFAS). Restraint systems are generally preferred to arrest systems for several reasons—most importantly because a fall is not possible with a restraint system—and are recommended whenever practicable.

137. D: A near miss indicates that something went wrong at the worksite, and if human error was involved, the safety professional should schedule competency assessments for all relevant parties. Insufficient knowledge and inadequate skills are leading causes of accidents, and competency assessments can determine whether the workers require additional training. Thus, Choice *D* is the correct answer. Admonishing the manager would likely be unproductive, and it could seriously damage the group dynamics. Therefore, Choice *A* is incorrect. Given the limited facts provided in the question, a targeted training program would likely be more appropriate than a generic, company-wide lecture. So, Choice *B* is incorrect. Likewise, providing the workers involved in the near-miss accident with instructional materials could be useful, but a competency assessment could provide a more accurate evaluation as to whether additional training is necessary. Therefore, Choice *C* is incorrect.

138. A: Stormwater is generated when rainwater hits the ground, accumulates contaminants, and then reenters a body of water. Thus, Choice *A* is the correct answer. Ephemeral surface water is a body of water that exists on a seasonal basis, such as a swamp or lagoon. Therefore, Choice *B* is incorrect. Stormwater has the potential to contaminate groundwater and render it toxic, but stormwater itself is not groundwater. In other words, Choice *A* provides an accurate definition of stormwater, while Choice *C* describes a potential consequence of stormwater. Therefore, Choice *C* is incorrect. Untreated wastewater is hazardous to the environment, but it is not directly related to stormwater. So, Choice *D* is incorrect.

139. C: Choice *C* is correct because practical demonstrations are preferred, as they allow employees to demonstrate their relevant skills and knowledge. Choice *A* is incorrect because written tests may be used to assess knowledge, not workplace conduct. Choice *B* is incorrect because verbal tests may show theoretical knowledge but not practical skill. Choice *D* is incorrect because nuanced demonstration is not a type of competency assessment.

140. A: Choice *A* is correct because the code of ethics encourages safety professionals to conduct their professional duties and activities with honesty and integrity. Choice *B* is incorrect because legal liability is an obligation imposed on a person or artificial person that resulted from a harm that was caused to another. Choice *C* is incorrect because a regulatory compliance manual would govern formal professional standards in a company's workplace. Choice *D* is incorrect because safety assessments may be performed by companies to determine workplace safety.

141. B: If the worker is stressed at 20 minutes, the worker must take a 40-minute break every hour. The temperature range is between 88 and 90 °F. The average of these two temperatures, or midpoint, is $\frac{88 + 90}{2} = 89\ °\text{F}$.

142. C: Information gathered with surveys can be used to determine the strengths and weaknesses of the culture, policies, and programs. Compilation of scores and the deviation of averages indicate the elements that can be addressed by process or department.

143. B: In chemical equations, the reactants are conventionally on the left side of the arrow. The direction of the reaction is in the direction of the arrow, although sometimes reactions will be shown with arrows in both directions, meaning the reaction is reversible. The reactants are on the left, and the products of the reaction are on the right side of the arrow. Chemical equations indicate atomic and molecular bond formations, rearrangements, and dissolutions. The numbers in front of the elements are called coefficients, and they designate the number of moles of that element accounted for in the reaction. The subscript numbers tell how many atoms of that element are in the molecule, with the number "1" being understood. In H_2O, for example, there are two atoms of hydrogen bound to one atom of oxygen. The ionic charge of an element is shown in superscripts and can be either positive or negative.

144. C: Training topics must include identification and knowledge of blood-borne pathogen hazards, control methods, emergency actions, personal protective equipment (PPE), the hepatitis B vaccine, the employer's exposure control plan, and general housekeeping considerations. Basic first aid, cardiopulmonary resuscitation (CPR), and automated external defibrillator (AED) are not topics addressed by the blood-borne pathogen standard, though personnel tasked with duties involving these topics would typically be required to receive blood-borne pathogens training. First-aid training is acceptable to OSHA if it is equivalent to Red Cross or Bureau of Mines curricula and is readily available in many locations; no such provision applies to blood-borne pathogens training. OSHA's blood-borne pathogens standard is found in 1910, subpart Z, Toxic and Hazardous Substances.

145. D: According to OSHA, the top three most common workplace safety violations are fall protection in construction, hazard communication, and scaffolding (wearing hardhats). A fleet driver must be familiar with hazardous materials and carry and review safety data sheets. Specifically, fleet drivers must understand how hazardous materials should be safely transferred and be familiar with chemical names.

146. B: Engineering controls are those that directly place a barrier between a worker and a hazard or remove the hazard completely. Examples include ventilation, shields, and noise absorbers. These are considered the most effective mitigation and reduction strategies for hazards. Personal protective equipment, such as gloves and goggles, provide barriers from hazards directly on a person's body. They are typically effective but more vulnerable to being breached or to user application error, so Choice *A* is incorrect. Administrative and work practice controls are trainings, process flows, and other policies that are eliminated to reduce risks, such as worker safety education. However, these are always susceptible to user error. Administrative and work practice controls are considered more effective than personal protective equipment but less effective than engineering controls; therefore, Choices *C* and *D* are incorrect.

147. A: Choice *A* is correct because ISO 14001 establishes a framework that allows organizations to manage environmental effects. Choice *B* is incorrect because OHSAS 18001 is not an ISO standard, and it is similar to ISO 45001 in regards to workplace safety. Choice *C* is incorrect because ISO 19011 concerns an organization's auditing procedures, not environmental effects. Choice *D* is incorrect because ISO 45001 focuses on workplace safety and risk reduction in the workplace.

148. B: Administrative, engineering, and substitution controls are elements of the control hierarchy, whereas operative is not.

149. C: The results of a BBS in the worksite—if successful—will cause a decrease, not increase, in worksite incidents. Direct contact between the workers and OSHA is not necessarily associated with a successful BBS.

150. C: When a die is rolled, each outcome is equally likely. Since it has six sides, each outcome has a probability of $\frac{1}{6}$. The chance of a 1 or a 2 is therefore $\frac{1}{6}+\frac{1}{6}=\frac{1}{3}$.

151. C: Combustible dust explosions often occur without warning. Dust particulates build up in a space, and pressure increases as they react with one another. This process usually is not visible, may not have a smell, and may not provide any other indicators until the explosion occurs. They typically occur in tightly confined spaces, such as factory environments without proper ventilation; therefore, these explosions cannot set off natural disasters that have to do with dust nor pollute large freshwater areas, so Choices *A* and *D* are incorrect. Combustible dust explosions can be prevented with safety processes that mitigate the five necessary elements (combustible dust use, heat, oxygen, confinement, and dispersion); this is an important aspect of workplace safety, so Choice *B* is incorrect.

152. A: Both the back-up alarm button and the emergency stop are located on the forklift and may be confused for one another, and they provide different functions in different contexts. An emergency stop when a forklift is trying to reverse can be inconvenient and possibly harmful, while a back-up alarm may not help when a worker needs to stop the equipment quickly. A vehicle not starting unless its key fob is in the car is a safety feature found in many newer cars, so Choice *B* can be eliminated. Music during a work shift is not necessarily harmful; there is no indication this would compromise worker output, so Choice *C* can be eliminated. Using the same model of refrigerator for chemicals and food would only be harmful if they were located next to one another or in the same room. There is no indicator that the two refrigerators are near each other, so Choice *D* can be eliminated.

153. B: Parasites are commonly found in drinking water sources, especially in less developed areas where water sources are not treated. Due to the timing of when Ben's symptoms started in relation to when he had the local drink, it seems most likely that the drink contained contaminated water. Jet lag does not usually result in such severe gastrointestinal distress, nor would it be likely to result during the return flight, so Choice *A* can be eliminated. While heat exhaustion may be characterized by nausea and sweating, it is not usually accompanied by diarrhea or abdominal pain. The question states it was a warm day but does not indicate there was excessive or insufferable heat. There are enough conflicting factors that Choice *C* can be eliminated. Fungi can occasionally contaminate drinking water, but it is not as common as parasites, especially in tropical locations. A parasite is more likely to be causing this infection; therefore, Choice *D* can be eliminated.

154. C: Competency assessments are an evaluative tool, and they're most often used in the workplace to determine whether a worker has sufficient skill or knowledge. Additionally, when analyzed cumulatively, competency assessments can provide quantitative data that's useful for evaluating the training program. For example, if numerous workers fail to acquire a certain skill, the training program likely needs to be adjusted. Thus, Choice *C* is the correct answer. Competency assessments typically function as the final, evaluative step of a training program, not as a learning tool, so Choice *A* is incorrect. Competency assessments don't necessarily involve hands-on learning, and in any event, that wouldn't be the primary reason they're used. Therefore, Choice *B* is incorrect. Choice *D* is incorrect because competency assessments are used at the end, not the beginning, of training.

155. C: Most chemical fires are a result of a primary or secondary chemical explosion. Chemical fires typically result from reactions that increase pressure, which causes an explosive reaction that causes heat and, ultimately, a fire. The primary or secondary explosions generally send shockwaves as well as particulates and projectiles that result in additional injuries. Electrical fires and downed trees are usually unrelated to chemical fires, so Choices *A* and *B* can be eliminated. Tinnitus is an ear condition in which a person hears intermittent ringing in the ears. It may develop later as an indirect result of being exposed to an explosion, but it is not directly from the chemical fire itself. Therefore, Choice *D* can be eliminated.

156. B: A monitor that can adjust in height allows the worker to place the monitor at eye level or lower, which supports neck and shoulder health. Monitors that are too high can place unnecessary tension on the cervical spine. When considering the best ergonomic practices for an office chair, whether it rolls isn't as important as neutral spine, hip, and foot posture in the chair. Therefore, Choice *A* can be eliminated. Corded desktop phones often hinder worker health and productivity by not allowing for hands-free work unless the worker cradles the phone between the ear and shoulder, which causes unnecessary neck strain, so Choice *C* can be eliminated. Laptops are convenient but do not provide ergonomic support on their own; ergonomic practices for eye health, spinal health, and arm positioning must still be implemented. Choice *D* can be eliminated.

157. A: This situation represents one of the elements of a permit-required confined space. With regard to entanglement and entrapment among all confined space types, these risks can never be completely eliminated. Each situation requires an analysis unique to the individual space. The introduction of equipment without analysis can exacerbate risk relative to inherent hazards, such as space orientation, method of entry/exit, and the presence of piping and other fixed or moving parts. Acquiring a permit takes an array of hazard elements into account, where feasibility of equipment and various precautions and procedures are analyzed before they're prescribed. Without the elements provided by a permit, there is not enough information to support either Choice *B* or *C*.

158. D: Choice *D* is an example of an organizational policy that keeps flames and burning ash away from equipment or materials that could start a fire. It also supports human health in other ways, such as eliminating particulates and secondhand smoke in the air. Choice *A* is incorrect because any electrical equipment can still start a fire if it is not properly used or maintained; only having electrical appliances in a break room does not necessarily reduce workplace fire hazards. Choice *B* and *C* are incorrect because having staff members wear personal protective equipment does not reduce overall workplace fire hazards, although it could protect the staff members from dangerous situations.

159. D: Something with high efficacy means it's likely to produce the intended result. On-the-job training has high efficacy because workers are trained in a real-world context, including the use of the same equipment, tools, and processes they will use every day to perform tasks. Thus, Choice *D* is the correct answer. On-the-job training might sometimes incorporate simulations, but on-the-job training usually centers on using the same equipment, tools, and processes used during the job. Therefore, Choice *A* is incorrect. On-the-job training doesn't emphasize independent learning; often, a mentor, coworker, or manager facilitates on-the-job training. Therefore, Choice *B* is incorrect. While written resources, Choice *C*, might be part of on-the-job training, the resources would most likely be used to support the performance of everyday tasks, which is the focus of on-the-job training.

160. A: Choice *A* is correct because the purpose of a gap analysis is to obtain desired performance. Choice *B* is incorrect because sufficient safety standards may be subject to a gap analysis, but they are not the purpose of a gap analysis. Choice *C* is incorrect because a gap analysis is not specific to crisis management. Choice *D* is incorrect because health conditions may be evaluated under a gap analysis, but they are not the purpose of a gap analysis.

161. A: Fire blankets cut off oxygen in the area. Fires require oxygen to keep burning, so by removing this element, a fire can be extinguished. Fire blankets do not create barriers between people and fires, so Choices *B* and *C* can be eliminated. They also do not remove fuel sources (e.g., gasoline, wood), so Choice *D* can be eliminated.

162. C: OSHA permits the use of a personal fall restraint system (PFRS) utilizing a body belt when a fall from an aerial lift is not possible, though this practice is generally discouraged in favor of a personal fall arrest system (PFAS). PFAS components are stronger and more widely available than those used for restraint and should be seen as providing a higher degree of safety and reliability. Another consideration when introducing body belts, even when properly used in restraint situations, is that employees may see them being used, incorrectly assume that they are acceptable in fall arrest situations, and utilize them as such. OSHA does not permit the use of guardrails as a form of fall protection for aerial lifts.

A suggestion to require that employees working over water in an aerial lift wear a fall or restraint harness would not be well advised because the drowning hazard may be greater than the fall hazard. When correcting hazards in construction, a safety professional must avoid the introduction of even greater hazards in the process and should always consider all options available to minimize risk. OSHA allows up to a six-foot drop in a PFAS, and this should always be minimized since an arrested fall can still cause injuries to the back or other body parts. When calculating the maximum arrest distance, the total length of all PFAS components must be addressed, not just the lanyard.

163. C: A job safety analysis (JSA) lists job steps and associated hazards and suggests corrective actions and procedures. A JSA can be developed before the start of a new task or be reviewed and revised as a reaction to an incident.

164. A: Entry and exit sign-in documentation prevents unauthorized and untrained people from entering potentially hazardous facilities. Choices *B, C,* and *D* focus on methods that keep an individual worker, rather than the public, safer from risks. Therefore, these choices are incorrect.

165. B: Only pairs I (harnesses and lanyards) and III (D-Rings and anchor points) are associated with the goal of fall protection. Fall protection is considered an engineering control and PPE. As a reactant to a hazardous energy release (the fall and gravity), the goal is to protect workers by preventing them from hitting the ground. Fall prevention equipment such as guardrails and restraints are associated with the goal of preventing the fall from occurring.

166. C: A fomite is a surface that might be contaminated with an infectious agent. In this case, if someone touches a door that has an infectious agent on it and touches their eyes, nose, or mouth, they could pick up the infection from the doorknob. Getting directly coughed on by someone who has a known infection, Choice *A*, is an example of infectious spread through droplets. Someone who did not wash their hands could spread infectious agents to food they prepared, which would then be consumed. This is infection through the oral route, so Choice *B* is not the correct answer. Herpes simplex type 2 is most commonly spread through direct sexual contact, not by sitting in the same room, so Choice *D* is incorrect.

167. B: Facilities that utilize processes to make different chemicals, such as lawn fertilizers, are at a higher risk for chemical fires due to the reactivity of the different materials that are used. A residential home with an electric cooktop may be at a slight risk for an electrical fire but not a chemical fire unless the resident made poor choices while cooking. However, Choice *A* does not provide enough information to determine this, so Choice *A* can be eliminated. A manufacturing facility that engages in mechanical manufacturing may be at risk for other types of fires, such as electrical or heat, but chemical fire risk is relatively low. The same can be said for a studio that is involved in sound engineering. Therefore, Choices *C* and *D* can be eliminated.

168. A: Sand and dirt are the best choice for extinguishing a Class D fire. Sand and dirt act as a non-reactive powder. Dry powder fire extinguishers (Class D) are designed for reactive Group IA and IIA metals of the periodic table and work by separating the fuel from the air, which contains oxygen. Choice *B* is incorrect because the use of water is dangerous and will result in a vigorous or violent reaction when in contact with Class D metals. Choice *C* is incorrect; reducing the oxygen content in an open environment is not practical. Choice *D* is incorrect because a cotton blanket is flammable.

169. B: Choice *B* is correct because candles are likely not proper response equipment; flashlights would be better because of their durability and reliability. Choice *A* is incorrect because an eye wash station may help remove debris and other materials from a person's eye in a crisis. Choice *C* is incorrect because batteries would be helpful in the event of a potential electrical outage. Choice *D* is incorrect because a plan would be a substantial tool in the event of an immediate crisis response.

170. D: Neither heat rash, Choice *A*, nor heat stroke, Choice *B*, cause permanent tissue damage. Heat rash is caused by clogged sweat glands and can heal without permanent damage. Heat stroke is characterized by a headache, vomiting, dizziness, and, eventually, passing out.

171. D: The presence of water is among the greatest hazards in excavation trenches. Threat of fissuring and collapse is magnified in less cohesive soils. Hazards are exacerbated with increased depth as lateral pressure increases where the soil can't support itself.

172. A: A cost-benefit analysis is an objective empirical study of the precise effects of a specific policy or plan. Cost-benefit analyses are critically important because they indicate if a policy or plan will save resources or squander them. If the costs outweigh the benefits, then an action isn't financially sensible. But if the analysis indicates that benefits will outweigh costs, then the policy can be pursued with confidence.

173. C: Leading and lagging indicators both include KPIs that help an organization evaluate its performance relative to strategic planning objectives and make adjustments to optimize future performance.

174. C: Choice *C* is correct because Rachel is able to work, but she cannot complete all of the required duties. Choice *A* is incorrect because the facts did not state that Rachel was unable to perform all of her required duties. Choice *B* is incorrect because the facts did not state whether Rachel will perpetually suffer from the injury. Choice *D* is incorrect because the facts did not state the duration of Rachel's injury or whether she is expected to fully recover.

175. C: First, calculate the TWA using the TWA formula below with D = 20%.

$$TWA = 16.61 \times log_{10}\left(\frac{20}{100}\right) + 90 = 78.39\ dBA$$

Now, calculate the estimated exposure:

$$Estimated\ exposure\ (dBA) = TWA\ (dBA) - (NRR - 7) = 78.39 - (25 - 7) = 60.39\ dBA$$

176. C: The purpose of the chain of command is to create an efficient response from a special and streamlined group of individuals during an emergency. The level of title and years of experience (under normal conditions) are not factors for performance and efficiency in responding to emergency situations. Job responsibilities are also not a factor unless they would preclude the individual from being available to serve in emergency situations.

177. C: BCSP requires certified professionals to conduct relationships with integrity to avoid conflicts of interest and report misconduct from any certified professional. Choices *A*, *B*, and *D* are all included in the BCSP code of ethics but do not relate directly to conflicts of interest.

178. D: The only exception to the requirement to provide a continuous path to ground in construction is for tools that are double insulated. Double-insulated tools are indicated by the symbol of a square inside another square.

179. C: Choice *C* is correct because ethical principles require an objective and honest representation of content. Choice *A* is incorrect because preferential sampling can create a misrepresentation in content. Choice *B* is incorrect because ethical principles would require a broad sampling to show greater objectivity. Choice *D* is incorrect because exaggerated sampling would create misrepresentation and would not provide honest or proportional results.

180. D: A morning huddle, a brief team meeting with present staff members before the workday begins, is a good time to discuss any possible exposure risks that could occur during the day's processes and efficiently communicate with all staff members. These informal check-ins can help staff members identify possible hazards early. Choice *A* is incorrect because scheduling an external exposure audit is useful and necessary, but it is not an informal and regular method of assessment. Choice *B* is incorrect because asking an administrator about any reported exposures may provide some information, but it does not provide any information about current risks, nor does it communicate or address current risks with the day's staff. Choice *C* is incorrect because creating a process workflow is one step in an exposure assessment but does not constitute conducting an actual assessment.

181. D: Many materials are flammable, and entire structures are often made of flammable materials (e.g., wood). Items that are simply present in an area can keep fires burning. Therefore, removing fuel for the fire is the most difficult method of extinguishing a fire, and extinguishing mechanisms try to address other elements first. It is much easier to stop the exothermic reaction with chemical extinguishers or by suffocating the oxygen source, so Choice *C* can be eliminated. Multi-head sprinkler sources are not obsolete and are often the first safety defense in many buildings. Choice *A* can be eliminated. With modern devices such as sensitive smoke detectors and automated alert systems, emergency services can often reach fires in a timely manner. Choice *B* can be eliminated.

182. C: Imminent danger, catastrophes, fatal accidents, and complaints/referrals constitute OSHA methods of inspection, although supplemental inspections are not included in this protocol. Imminent danger means the employer will be asked to immediately remove workers from the worksite, as there are hazards that require instant corrective action and cannot be mitigated through normal channels.

183. A: Online surveys can be quickly distributed to a large group of people, can collect any type of information the researcher needs, and do not have paper waste. However, since surveys can have low completion rates, Mina may need to send regular follow-up reminders or ask leadership to make the survey required in order to reach her completion goal.

184. C: Choice *C* is correct because the audiometric test requires that an employee not be subject to excessive noise for 14 hours prior to the test. Choices *A*, *B*, and *D* are incorrect because these do not represent the amount of time an employee should avoid exposure to excessive noise before testing.

185. C: Administrative controls include worker selection and training, isolation barriers to control worker movements and access, and signage to caution or warn workers to increase awareness and safe behaviors. Substitution controls are used as redesigns and replacements of material, chemicals, and processes. Engineering includes the modification of materials and equipment to reduce hazard risk, while PPE is the equipment workers use to protect against hazards and is not a type of control.

186. A: Interpersonal communication requires communicators to send, hear, and respond to messages. Active listening is extremely beneficial because it demonstrates comprehension and improves the quality of feedback. Thus, Choice *A* is the correct answer. Choice *B* is incorrect because commercial expertise isn't necessary or directly related to listening. If anything, technical jargon can frustrate interpersonal communication. Likewise, Choice *C* is incorrect because experience with active listening and interpersonal communication is more directly relevant than general professional experience. Choice *D* is incorrect because interpersonal communication is much more closely associated with face-to-face communication than digital communication.

187. C: Per 1926.100, OSHA adopts the hard hat criteria of the American National Standards Institute (ANSI). Additional elements (e.g., expiration dates and/or life of service, hat parts and insert replacements) lie in the specifications of the manufacturer. ASTM involves material and product properties and performance. NIOSH is primarily a research and recommending entity. UL is a safety and quality testing entity traditionally associated with electrical components.

188. D: Choice *D* is correct because *ceiling limit* is the term for the exposure limit that should not be exceeded at any time. Choices *A*, *B*, and *C* are incorrect because these are not actual terms related to exposure limitations.

189. B: Contaminated food sources are one of the most likely causes for infection, as improperly stored or prepared foods can harbor viruses, bacteria, parasites, and fungus alike. Soil may contain bacteria and parasites, but it's not as common of a source as food since most people come into contact with food multiple times per day; therefore, Choice *A* is incorrect. Cleanrooms are generally sterile environments that are controlled to have low levels of contaminants, so they are unlikely to be a common source of infection; therefore, Choice *C* is incorrect. Anterooms are the changing areas for workers entering sterile environments. While these rooms may have significantly more contaminants than a cleanroom, they are a storage area and not a typical source of infection, so Choice *D* is incorrect.

190. C: An executive summary is condensed for the reader's convenience. Readers should have all the important information available, without the burden of poring through multiple pages and details.

191. C: In an emergency, the first priority is to safeguard personnel, equipment, and processes, in that order. Relevant emergency management personnel should be advised of the situation as soon as possible, but ensuring that personnel are safe is always the most important consideration during an emergency. OSHA's requirements for emergency action plans (EAPs) are outlined in 1926, subpart C, General Safety and Health Provisions, and also 1910, subpart E, Means of Egress. OSHA's Hazardous Waste Operations and Emergency Response requirements are found in 1910, subpart H, Hazardous Materials. There is no requirement that alarms must be 105 A-weighted decibels, but any alarms, when used, should be loud enough that they will be heard by all employees who will need to hear them.

192. C: Choice *C* is correct because meetings are scheduled events designed for a specific topic for a group setting. Choice *A* is incorrect because a speech is intended to convey a message to a larger group. A group of 10 recipients is likely considered a moderate size depending on the company. Choice *B* is incorrect because verbal communication has a greater impact and provides clearer communication than written communication since it allows recipients to ask additional questions for better understanding. Choice *D* is incorrect because a phone conference may be subject to technical failures and different background noises.

193. C: A diverse workforce can include personnel of different nationalities, which presents barriers to written and oral forms of communication. Signage uses universally accepted symbols and pictograms. An example of this widespread use is the Global Harmonization System of Classification and Labeling being implemented in many nations.

194. B: On a modern construction site, *dry chemical* and *multipurpose dry chemical* extinguishers are the same, and both phrases are used. Prior to the development of modern "multipurpose" dry chemical extinguishing agents, which are suitable for Class A, B, and C fires, the first dry chemical extinguishers were suitable only for Class B and C fires and have become unpopular. Fire extinguishers on construction sites should be inspected at least monthly; this is an NFPA (National Fire Protection Association) consensus standard that OSHA has adopted and enforces. An extinguisher must be located within 50 feet of five or more gallons of flammable materials in an outdoor construction environment, not 10 feet. There is no mandatory fire extinguisher to employee ratio, such as one extinguisher for every 10 employees.

195. C: The manipulation of powdered materials can be challenging and dangerous because the powder is likely to escape segregation and separation. Glove box systems limit this risk because it isolates and contains powders during manipulation. Thus, Choice *C* is the correct answer. Barricades, Choice *A*, and locked cabinets, Choice *B*, are used to increase physical security. Although physical security is an important aspect of segregation and separation, neither would specifically assist in the manipulation of powdered materials. Therefore, Choices *A* and *B* are both incorrect. Similarly, sealed systems can be used to segregate materials, but this type of system is more often used to control and limit the inhalation of dust. Therefore, Choice *D* is incorrect.

196. C: Thermal detectors are accurate and inexpensive sensors that primarily work best in specific contexts. A thermal detector would be useful near a home's natural gas pipe because it would be able to sense the heat from a leak. A natural gas leak is especially dangerous, as it is noxious to residents and highly flammable. A thermal detector would be able to note a change in the temperature around the pipe if there was a leak, which a smoke detector or fire detector may not be able to do. A thermal detector would not be useful near a window, which is vulnerable to high temperature changes based on the position of the Sun, so Choice *A* can be eliminated. Thermal detectors are also not useful fire safety mechanisms in large buildings, as they cannot detect changes in heat quickly enough in large spaces. Therefore, Choices *B* and *D* can be eliminated.

197. C: Risk matrices use event frequency vs. severity to determine risk level either quantitatively or qualitatively. Failure mode and effect analysis applies mathematical modeling to determine potential rates of failure and reliability for systems such as materials and equipment. "Why?" methodology is used in incident investigations to determine root cause. Job safety analysis is used to detect hazards associated with job steps and apply corrective actions for each hazard.

198. B: Ergonomics takes into consideration that workers are anthropometrically diverse. Specifically, a person's hand size and shape dictate how power hand tools are used. As the absent production workers appear to be physically different than Janine (one is considerably larger and more muscular, and the other is considerably younger), it is likely she may require a different duration of use for the hand tool portion of the production line. However, she should be able to still complete these tasks, including lifting items. There is no indication in the question that she would not be able to complete the task at all. Therefore, Choices *A* and *D* can be eliminated. Since there are three workers offering to support during the technicians' absences, it seems unnecessary to hire new staff for this period of time, so Choice *C* is incorrect.

199. C: Choice *C* is correct because ethical writing should be unambiguous (clear) and objective. Choice *A* is incorrect because writing must be an objective perception, not a subjective perception, and it should be competent. Choice *B* is incorrect because writing should be unbiased and honest. Choice *D* is incorrect because statements and opinions should be clear and proportionate to material facts.

200. C: When working with hot tools, sparks can fly up to 35 feet and ignite flammable gas in the air or get caught on clothing or in cracks in the walls and floors.

Index

Accelerated Silicosis, 126
Acceleration, 32, 33, 34, 35, 63, 64, 93, 201
Access Controls, 122
Active, 63, 73, 76, 83, 96, 116, 153, 174, 182, 189, 192, 220
Acute Exposures, 127
Acute Silicosis, 126
Additive Effect, 128
Administrative, 38, 39, 50, 51, 53, 76, 78, 94, 136, 162, 164, 171, 185, 186, 192, 195, 215, 220
Alpha, 77, 78, 130, 131, 169, 199
American Council of Governmental Industrial Hygienists, 117
Antagonistic Effect, 128
Anthropometry, 98, 166, 197
Arc Flash, 101
Armed, 83, 201
As Low As Reasonably Practicable (ALARP), 95
Assessment of Chemical Hazards, 121
Atomic Mass, 23, 203
Atomic Weight, 23, 24, 173, 204
Attendant, 69, 167, 198
Audits, 48, 88, 140, 145, 149, 156, 157, 159, 202
Automatic Braking Systems, 63
Bacteria, 27, 117, 118, 134, 135, 221
Behavioral Factors, 91
Biological Hazards, 117, 134, 204, 209
Bomb Threats, 116, 201, 212
Budget Matrix, 90
Carcinogen, 82, 121, 124, 209
Cause and Effect Diagrams, 58
Caution, 65, 71, 84, 87, 147, 220
Ceiling Limit, 124, 192, 221
Change Analysis, 60
Checklist Analysis, 59
Chemical Absorption, 127
Chemical Hazards, 83, 100, 117, 121, 122, 126, 134, 169, 199, 200, 204, 209
Chemical Ingestion, 127
Chemical Injection, 126, 127
Chemical Reactions, 12, 13, 22, 105
Chemical Solutions, 14
Chronic Exposures, 127
Circuit, 14, 15, 16, 18, 19, 20, 21, 66, 101, 181, 209, 210
Civil Liability, 155, 181, 211
Close-Call Reporting, 49
Coefficient, 12, 36
Combustible Dust Fire Pentagon, 105
Confined Space (CS), 69
Conservation, 12, 20, 36, 38, 92, 134, 135, 137, 138, 164, 173, 174, 195, 204
Conservation of Charge, 20
Continuous, 48, 49, 69, 76, 88, 89, 96, 103, 106, 150, 165, 172, 191, 207, 219
Contract, 155, 169, 181, 200, 211
Copyright, 158
Cost Management, 90
Cost of Risk, 37
Criminal Liability, 155, 181, 211
Current, 14, 15, 17, 18, 19, 20, 21, 22, 35, 79, 80, 83, 100, 101, 102, 103, 139, 152, 153, 156, 157, 158, 161, 162, 163, 165, 174, 178, 197, 201, 205, 206, 208, 220
Danger, 65, 76, 91, 113, 114, 115, 134, 135, 141, 191, 210, 220
Days Away from Work Cases, 47
Defensive Driving, 62, 72, 185
Descriptive Statistics, 39
Direct Costs, 37, 38, 48
Distracted Driving, 62
Dormant, 83
Dose Level, 24
Dry Chemical, 107, 108, 109, 143, 193, 221
Early Symptom Intervention, 97
Electric Circuit, 21
Electric Surges, 101
Electrical Power, 18, 176
Electrical Resistance, 20
Electricity, 18, 19, 22, 100, 101, 105, 138, 170, 201, 209
Elimination, 50, 51, 52, 76, 78, 94, 134, 154
Energy, 18, 19, 20, 21, 35, 36, 37, 38, 60, 66, 67, 72, 81, 100, 101, 111, 130, 131, 138, 139,

164, 166, 170, 172, 175, 195, 197, 204, 206, 218
Energy Trace and Barrier (ETBS) Analysis, 60
Engineering, 27, 50, 53, 58, 59, 74, 76, 77, 78, 81, 82, 94, 114, 129, 136, 137, 162, 171, 176, 183, 185, 186, 189, 192, 206, 215, 218, 220
Entrant, 69, 187, 198
Environmental Management Systems Standards, 140
Ergonomic Stressors, 91, 92, 93
Ex-Situ Remediation, 143, 165
Extreme Weather Events, 114
Failure Mode and Effects Analysis (FMEA), 58, 82
Fall Hazards, 68, 70, 71, 84
Fall Protection, 31, 32, 62, 70, 71, 74, 81, 164, 170, 188, 195, 201, 215, 217, 218
Fatigue, 61, 63, 72, 91, 92, 93, 97, 98, 126, 205
Feedback, 76, 90, 94, 148, 150, 151, 153, 154, 164, 166, 167, 186, 197, 198, 199, 220
Fiber Per Cubic Centimeter (F/Cc), 124
Final Ruling on Crystalline Silica, 161
Fire Tetrahedron, 100, 105, 106
Fire Triangle, 100, 105
Flash Point, 100, 110, 165, 196
Flashover, 101
Force of Gravity, 34, 35
Forklifts, 72, 95
Friction, 20, 35, 36, 37, 91, 101, 111, 130, 178, 208, 209
Friction Factor, 36
Gap Analysis, 79, 80, 151, 153, 174, 188, 205, 217
Globally Harmonized System, 65
Goal-Setting Theory, 151
Half-Life, 22, 23, 173
Hazard, 11, 49, 50, 51, 52, 53, 54, 55, 56, 57, 60, 62, 65, 66, 69, 72, 75, 76, 77, 78, 79, 80, 82, 83, 84, 86, 88, 89, 94, 98, 102, 106, 109, 110, 112, 117, 120, 122, 124, 125, 126, 129, 130, 134, 140, 147, 149, 156, 157, 160, 161, 162, 169, 173, 179, 180, 185,188, 193, 195, 199, 201, 202, 209, 210, 215, 217, 218, 220, 222
Hazard Analysis, 56, 57, 76, 161
Hazard Communication, 65, 76, 147, 161, 162, 215
Hazard Controls, 50, 57, 79, 80, 82
Hazardous Waste, 134, 138, 141, 142, 143, 166, 168, 183, 195, 199, 204, 213, 221
Heat Stress, 29, 30
Heating, Ventilation, and Air Conditioning (HVAC), 27
Hierarchy of Conservation, 138
Hierarchy of Controls, 50, 51, 77, 84, 149, 171, 186
Higher-Level Controls, 51
Immediately Dangerous to Life and Health (IDLH), 124
Impulse, 35, 36
Indirect Costs, 37, 38, 48, 62
Industrial Hygiene, 117, 150, 156
Inhalation, 100, 121, 124, 125, 126, 127, 134, 169, 200, 222
In-Situ Remediation, 143, 169, 196, 200
Inspection Frequency, 48
International Organization for Standardization (ISO), 49, 140
International Union of Pure and Applied Chemistry (IUPAC), 11
Inverse Square Law, 25
Ionizing Radiation, 130, 131
Isaac Newton's Three Laws of Motion, 32
Jackknife, 64
Job Rotation, 96, 207
Job Transfer Or Restriction Cases, 47
Kurtosis, 41
Ladders, 32, 68, 70, 71, 74, 81, 164, 180, 198, 202, 209
Lagging Indicators, 46, 190, 219
Law of Conservation of Energy, 19, 20
Law of Conservation of Mass, 12
Leading Indicators, 48, 190, 219
Legal Liability, 155, 160, 185, 214
Life Cycle Costing (LCC), 38
Lighting, 61, 62, 88, 91, 92, 101, 175, 205
Load Capacity, 26, 74, 95, 147, 166, 183, 197
Lockout/Tagout (LOTO), 66
Low-Level Controls, 51
Machine Guarding, 53, 72
Management of Change, 55, 56, 76
Manmade Disaster, 114, 176, 201, 206
Mean, 19, 39, 40, 41, 42, 45, 121, 169, 172, 200
Mechanical Ventilation, 27, 110, 111
Median, 39, 40, 169, 200
Micrograms Per Cubic Meter (µg/M3), 123

Mid-Level Controls, 51
Milligrams Per Cubic Meter (Mg/M3), 123
Millirem, 131
Mitigating Risk, 53
Mixed-Mode Ventilation, 110
Mode, 39, 40, 57, 110, 146, 149, 169, 173, 193, 200
Mold, 117, 118, 119, 134, 180, 209, 210
Momentum, 33, 35, 36, 178
Multipurpose Dry Chemical, 108, 193, 221
Mutagen, 120, 177, 207, 213
National Institute for Occupational Safety and Health (NIOSH), 24, 76, 98, 117
Natural Disasters, 49, 53, 90, 114, 170, 186, 206, 207, 215
Natural Ventilation, 27, 110, 111, 164, 195
Near-Miss Or Near-Hit, 49
Needs Analysis, 151, 152, 153, 179, 184, 208, 213
Negligence, 54, 155, 161
Noise, 24, 28, 29, 76, 92, 117, 129, 130, 148, 153, 166, 174, 191, 196, 197, 208, 215, 220
Noise Dose, 28
Numeric, 77, 78
Ohms, 15, 18, 20, 21, 165
Oral Communication, 145, 148
Other Recordable Cases, 47
Outliers, 39, 40, 41, 169, 200
Outsourcing, 55
Parasites, 117, 118, 133, 216, 221
Parts Per Million, 24, 123, 124
Passive Restraining Systems, 63
Performance Management, 90
Personal Fall Arrest System (PFAS), 31
Personal Protective Equipment, 50, 53, 57, 61, 78, 105, 119, 122, 133, 162, 171, 185, 192, 215, 217
Plagiarism, 158
Potentiation Effect, 128, 129
Prefix, 11, 12
Preventative Costs, 37
Probability, 41, 42, 43, 44, 45, 46, 57, 77, 115, 122, 186, 215
Probability Distribution, 43
Products, 12, 13, 65, 90, 100, 124, 138, 139, 196, 205, 206, 214
Project Management, 77, 89, 90, 94
Psychological Influences, 91
Radiation, 23, 25, 26, 29, 30, 117, 128, 130, 131, 169, 199, 210
Radioactive Decay, 22, 23, 130
Radioactivity, 22, 23, 199
Range, 39, 41, 43, 45, 85, 99, 106, 139, 167, 214
Reactance, 18
Reactants, 12, 13, 185, 214
Recognize, Evaluate, and Control, 77
Recycling, 137, 138, 139, 174, 176, 206
Reduce, 32, 37, 51, 57, 58, 60, 62, 63, 64, 65, 74, 77, 78, 81, 83, 84, 88, 91, 92, 93, 95, 96, 97, 103, 110, 111, 112, 115, 120, 126, 128, 129, 130, 136, 138, 139, 144, 156, 160, 164, 165, 169, 174, 176, 178, 187, 197, 198, 200, 202, 204, 205, 207, 215, 217, 220
Repetitive Stress Injuries, 92
Resistance, 14, 15, 17, 18, 19, 20, 21, 33, 56, 165, 202, 208, 209
Restraining Systems, 63
Reuse, 138, 139, 174
Rigging Hitch, 26
Risk Management, 49, 53, 54, 85, 90, 132, 140, 160
Safety Interventions, 49
Same-Level Falls, 70
Sampling, 83, 143, 158, 191, 220
Scaffolding, 31, 70, 73, 96, 215
Scheduled, 76, 89, 149, 221
Scheduling, 80, 84, 90, 154, 199, 220
Section 5(a) (1) General Duty Clause, 93
Segregation, 81, 110, 222
Separation, 81, 82, 110, 138, 147, 203, 222
Shooting, 115, 181, 182, 210
Signs and Symbols, 147
Silica, 124, 125, 126, 162, 177, 181, 207, 208
Skidding, 63, 64
Sling, 26, 27, 183, 213
Solid Waste, 137, 138, 143, 176, 206
Solvent, 14
Source Strength, 23, 24, 173, 203, 204
Specify, 24, 71, 197
Sprinkler Types, 107, 109
Standard Deviation, 41, 42
Standard Writing, 158
Static Electricity, 101, 179, 209
Stormwater, 135, 177, 184, 207, 214
Subscript, 12, 18, 214

Substitution, 13, 18, 50, 52, 76, 78, 94, 171, 186, 192, 197, 215, 220
Suffix, 11, 12
Sustainability, 137, 138, 195
Swing-Out, 63, 64
Symptoms, 63, 69, 93, 118, 120, 122, 123, 126, 127, 179, 208, 216
Synergistic Effect, 128
Target Organs, 123
Teratogen, 121, 213
the Department of Labor (DOL), 72
Time-Weighted Average (TWA), 202
Tons Per Square Foot (t/sf), 67
Tort, 155, 169, 181, 200, 211
Toxicity, 24, 82, 121, 122, 124, 134, 168, 179, 199, 209, 211
Trailer Sway, 64
Training Frequency, 49
Trucks, 62, 72, 96
Tubular Welded Frame, 73
Ultraviolet (UV), 131
Uniform Probability Distribution, 43, 44
Universal Precautions, 119, 172, 203
Value Engineering, 54
Vandalism, 116, 175, 205
Ventilation, 27, 28, 53, 69, 110, 111, 126, 127, 136, 164, 168, 181, 183, 199, 211, 215
Verbal Threats, 116, 201
Vermin, 134, 173, 204
Viruses, 115, 117, 118, 134, 211, 221
Voltage, 14, 15, 17, 18, 19, 21, 81, 101, 197
Vroom Expectancy Theory, 151
Warning, 63, 65, 67, 82, 168, 186, 215
Wastewater, 135, 168, 184, 199, 211, 213, 214
What-If Method, 59
What-If/Checklist Analysis, 59
Wind Chill, 30
Work Hardening, 96, 169, 199
Workplaces, 53, 100, 108, 115, 117, 132, 153
Work-Related Musculoskeletal Disorders (WMSD), 95
Written Communication, 145, 148, 177, 192, 221
Z-10 Risk Assessment Matrix, 78

Dear ASP Test Taker,

We would like to start by thanking you for purchasing this study guide for your ASP exam. We hope that we exceeded your expectations.

Our goal in creating this study guide was to cover all of the topics that you will see on the test. We also strove to make our practice questions as similar as possible to what you will encounter on test day. With that being said, if you found something that you feel was not up to your standards, please send us an email and let us know.

We would also like to let you know about other books in our catalog that may interest you.

CHST

amazon.com/dp/1628459085

We have study guides in a wide variety of fields. If the one you are looking for isn't listed above, then try searching for it on Amazon or send us an email.

Thanks Again and Happy Testing!
Product Development Team
info@studyguideteam.com

www.ingramcontent.com/pod-product-compliance
Lightning Source LLC
Chambersburg PA
CBHW080756150926
51501CB00045B/1247